Science Fair
Developing a Successful and Fun Project

Maxine Haren Iritz
Photographs by A. Frank Iritz

TAB TAB BOOKS
Blue Ridge Summit, PA

To Michael and Stuart, whose projects
inspired this book,
and
To all Science Fair participants
past, present, and future
Everywhere

FIRST EDITION
SIXTH PRINTING

© 1987 by **Maxine Iritz** and **A. Frank Iritz**.
Published by TAB Books.
TAB Books is a division of McGraw-Hill, Inc.

Library of Congress Cataloging-in-Publication Data

Iritz, Maxine Haren.
    Science fair.

    Bibliography: p.
    Includes index.
    1. Science—Exhibitions. I. Title.
    Q105.A1I75   1987   507′.8   87-10030
    ISBN 0-8306-0936-9
    ISBN 0-8306-2936-X (pbk.)

TAB Books offers software for sale. For information and a catalog, please contact
TAB Software Department, Blue Ridge Summit, PA 17294-0850.

Questions regarding the content of this book should be addressed to:

Reader Inquiry Branch
TAB Books
Blue Ridge Summit, PA 17294-0850

Cover photographs by Michael Bruce Iritz

Contents

Introduction

As you know, there is currently a great emphasis on raising our educational standards. Because our society is increasingly reliant on technology, the particular stress has been on math, science and computers. Therefore, school districts throughout the country are concentrating energy and resources in these fields.

The showplace for many of these efforts is the annual science and engineering fair, for which students prepare and exhibit science projects. These projects require students to conduct research and experimentation needed to prove a hypothesis. Once completed, these projects, together with displays, are entered in individual school fairs where winners from each category, or scientific field, advance to a regional science fair. The most outstanding projects proceed to state or international competitions, where recognition, as well as scholarships, trips and other prizes, can be considerable.

You're probably reading this book because you (or perhaps your son or daughter) are planning to do a science project, and are looking forward to competing in a science and engineering fair. Although educationally valuable and often emotionally satisfying as well, a science project requires a major commitment of time and effort. Along the way, students usually need not only encouragement and moral support, but some very concrete assistance, too.

Preparation can often take up to a year of intense effort, from the crucial first step, the selection of a topic appropriate to the interests and abilities of the student, to the construction of the display at the project's conclusion. Errors can be costly, both in terms of money spent for supplies and equipment, and the time and energy diverted from the ultimate goal.

Parents are most anxious to assist. They are, in fact, a crucial factor in the success of a project, especially for a first time participant. Unfortunately, many parents, who themselves never did a project and whose scientific background may be limited, often do not know how to help their youngsters. Despite considerable guidance, support and dedication from teachers and advisors, there are many facets of science fair participation which, to the first-timer, are formidable. A teacher may sometimes supervise over thirty projects a semester, and therefore cannot always deal with the day-to-day problems of conducting each student's experiment.

This book will present a clear guide on how to complete a successful science project. Our objective is to lead students and their parents through the steps involved, giving some helpful hints along the way. Teachers, judges, participants, and their families have all contributed their feelings, observations, and best advice. This is intended to guide you down the easiest road to success, and help you avoid the pifalls.

The appendices were designed to help your son or daughter with that important first step, selecting a topic. In Appendix B, a sampling of projects, we have not always included the top prize winners, but instead, projects from several disciplines, as well as varying degrees of complexity. In this manner, we have attempted to reach students' individual interests and of their particular degree of scientific knowledge. We've also included a glossary, summarizing the important terminology introduced throughout the book.

In speaking to science fair winners, the most successful participants had a great deal of help and support from their families. We'd like to help you and your youngster achieve not only success, but enjoyment and satisfaction from the entire science fair experience.

We wish to thank everyone whose help and expertise made this book possible. To those who managed the 37th International Science and Engineering Fair in Fort Worth, Texas, thank you for making our trip productive. To the members of The Greater San Diego Science and Engineering Fair Management committee and Student Advisory Board, most especially Philip D. Gay, Ronald E. Domb, Mary Domb Mikkelson, and Howard L. Weisbrod, and to San Diego science teachers Randy Borden of Bell Junior High School and Gary Oden of Morse High School, we owe a special thank you.

Finally, to all students whose projects are referenced or photographed here, muchas gracias, danke, merci, "dom arrigato," grazie, for helping to make this book a reality.

Thank you all.

Chapter 1

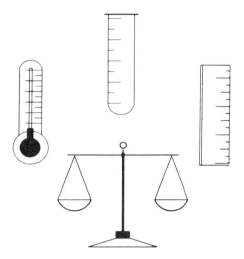

So You're Going to Do A Science Project!

Right now, your overwhelming feeling is probably panic. A science project seems like a huge undertaking, with many things to be done. You may be thinking that it will take so much time and work, you may never be able to goof off again! Therefore, the question uppermost in your mind may be *Why Do A Project?*

For some of you, the answer might be simply that it's required. If that's the case, the completed project will be a large portion of your science grade. Others might be doing the project because they are interested in some area of science and already have chosen their project idea. Another reason might be the sheer joy of competing in a science fair, matching your abilities against your peers.

For whatever reason you're beginning a science project, this is very likely your very first independent study. Working on your own is one of the most important benefits of doing a project. Although you'll have a great deal of help from parents, teachers and other resources, this project will be entirely your responsibility. You'll control the planning and scheduling as well as executing the research and experiment. Although you will use many of the procedures and techniques that you've learned in science lab, you'll have the freedom to set your own pace.

In the first phase of the project, you will select the subject for your experiment. This process gives you the opportunity to explore the general topics included in the various scientific disciplines. Perhaps, you'll learn about subjects you never thought about before.

Another crucial element is the background research paper. Throughout high school and college, you'll be judged on the ability to write a well researched and presented paper. In fact, for many of you, the research paper will be graded not only by your science teachers, but by your English teacher.

Completing a project will give each participant real "on the job training" in carrying out an experiment according to sound scientific methods. In your school labs, your teachers generally specify the steps to follow for each experiment and provide the necessary materials to work with. They are always on hand to insure that the procedure is carried out safely and properly. When completing your project, *you* must design your experiment, find the required materials, specify the procedures, safely carry out the experiment, and record and measure your observations, making sure that you properly adhere to scientific methods.

When the experiment is finished, you will need to compile and tabulate your results. To do this, you may need mathematical or statistical skills, as well as the ability to graphically show your results. Finally, you'll formulate your conclusions, which will show whether or not your experiment "worked." This will demonstrate your ability to correlate the facts you have researched with the results of your experiment.

In your concluding statement, you can also speculate on the possible benefits to society that your results might present. Here, you can also mention any plans you might have to continue working in that area. Sometimes, students become so fascinated with their subject area that year after year, they continue expanding on their research topic, until, by the time they're 12th graders, they've become experts in their field.

Doing a science project may satisfy the artist in you, too. Creating a display, which will illustrate and present a summary of your project, should show you and your work

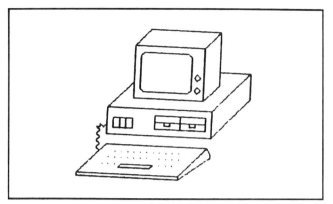

Fig. 1-1. Illustration of computer symbol.

to its best advantage. Producing an effective backboard will call on your design, artistic, and photographic skills.

If you've never competed in a science fair before, you might wonder what judges look for. From the school fair to the international level, judges agree that the steps and methods used in the experiment are usually more important than your results or whether you prove your theory.

For those of you who advance beyond your school fair, you'll probably get to spend a few days at the city, county, or state fair. There, guided tours to areas of interest are usually arranged. Professionals are sometimes available for counseling about scientific careers. Finally, you'll also get the chance to meet new and interesting people. In the process, you'll discover, within yourself, a poise and grace you never knew existed.

Although it is by no means necessary to have a home computer to do a successful science project, I have included *computer notes* throughout this book, shown with the symbol in Fig. 1-1. This will suggest the various ways to effectively use your hardware and software. For those of you who do not have your own system, several students said they were able to use their school computer labs after hours or on Saturday.

This book was written for you, the student, to make the science project experience the most enjoyable and educational experience possible. You may doubt your scientific abilities. This is only natural, since this is probably your first undertaking of this type.

Many other students, just like you, have started projects with the same mixture of uncertainty and fear. However, with a lively curiosity and a good measure of diligence they have created effective, scientifically acceptable, award winning projects.

For many of you, the learning will extend far beyond the areas of science to writing, art, mathematics and research. Along the way, I hope you will also find a true sense of pride and accomplishment, and learn not only about science, but about yourself.

Like any other long term effort, a science project is filled with many small victories, false starts, bursts of activity, periods of procrastination, sleepless nights and stomachs full

of butterflies. Through it all, regardless of awards or grades, every youngster who completes a project is a winner.

Remember that the main objective is learning. An exhibitor in the 1986 International Science and Engineering Fair, Stephen M. Jacobs, from Hamilton, Ontario, Canada designed and built a *Continuous/Pulsed Ion Beam Accelerator* (see Fig. 1-2). In the more than two years that it took to build and test the device, he "learned about the atom in every detail and from every angle. Everything that I learned with this project, every application I have identified and made known to others has given me much understanding of the atom and its infinite uses for man."

When Mr. Jacobs wrote about his project, he included a note to all young scientists. "Spending several years on a large scale science fair project may seem rather difficult to comprehend, but if you have a question that you would like answered (be it biological, behavioral, chemical, physical, etc.) the time you spend, answering it for yourself means that you have gained a greater amount of knowledge and experience—more so than if you were to refer to a text. Science has only one purpose—the full understanding of

Fig. 1-2. Stephen M. Jacobs; *Design, Construction, and Operation of a Continuous/Pulsed Ion Beam Accelerator.*

everything around us and within us. The projects you do in the future may well have some direct application for humanity. Perhaps you will find a cure for cancer, or maybe you could design some new form of space travel. Whatever you do, complete the project with the full realization that the time you have spent was spent well, and with every further question you ask, you are developing a greater understanding of who you are, where you are and what your purpose is here.'' Look at your science project not as an assignment but as a journey of discovery. Happy exploration!

Chapter 2

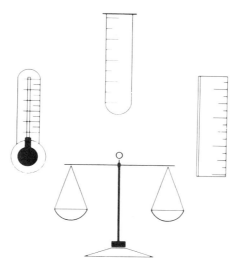

Choosing a Topic

If there is one factor that is critical to the success of all science projects, it is the choice of topic. In fact, some of the exhibitors, especially those competing for the first time, said that finding a subject was the most difficult part of the entire experience. Most students strove for an idea that was original, or at least, not overdone. Everyone wanted a concept that excited their curiosity, but they also wanted an experiment which, they hoped, would not be too difficult.

When looking for an idea, the most important consideration is finding a topic that is interesting and fun for you. Unlike other assignments or research papers you may have, a science project will take at least four months. You will need to do intensive research to gather background information and careful experimentation, in order to scientifically prove or disprove the theory that is the basis of the study. The documentation, analysis and presentation of your results and conclusions will also take considerable time. The work will be easier and far more fun if you're working on a subject you enjoy.

CATEGORIES

Many categories of projects compete in science fairs. They cover all scientific and engineering disciplines, ranging from botany to zoology, and everything in between. If you already have an idea of what your experiment will entail, it would be a good idea to see what category it falls into. If, on the other hand, you have no idea at all, examining the various categories might at least give you some insight into the general area you wish to work in.

A program from a past science fair contains a great deal of valuable information about the different categories. Usually, the projects competing in each category are listed, along with the name and school of each exhibitor. One of

the first things you may notice is that some categories have many more entries than others. This, especially at the junior division level, occurs for several reasons. One important reason is that these popular fields are covered in the junior high school science classes. Availability of background information in that particular category is another reason for having more projects in that area. Finally, in certain areas, the subject matter is more relevant to the students' lives, for example human behavior, computers, or medical science. However, these reasons, which make a certain area of study easier and more popular, will also mean stiffer competition in those categories.

Take special notice of concepts that have been presented very frequently. Your topic does not necessarily have to be original for your project to be a success, but teachers and judges will look for a creative approach. Therefore, if you plan to use an idea that has been done often, look at it from a fresh angle. Although information is more readily available in these areas, you may need to work harder to manage an original viewpoint.

Categories, as defined by the Greater San Diego Science and Engineering Fair, and some examples of projects in each grouping are:

☐ Animal Behavior
   —Interaction vs. Back-Up: Testing Two Theories of Mouse Navigation
   —Pitch Discrimination in Goldfish
☐ Biology/Microbiology
   —Bacterial-Fungal Synergism
   —Penicillin's Effect on Reproductive Rate of Volvox
☐ Botany
   —Effects of Ammonium Chloride and Potassium Nitrate on Spirulina

Fig. 2-1. David Calabrese; *TextLink.*

—Hydroponic vs. Soil Gardening
☐ Chemistry/Biochemistry
  —Biochemical Cycling by Microbial Life: The Nitrogen Cycle
  —Protein Electrophoresis Determination of Necessary Agents for Planaria Regeneration
☐ Computers
  —TextLink: The Printed Word to Computer via Structural Character Recognition.
  —Yosemite Island: A Computer Simulation of Evolution
☐ Earth/Space Science
  —Shape Analysis of Microfossils Using Laser Shadow
  —Effects of Sunspots and Time of Day on Radio Transmissions
☐ Engineering/Electronics
  —HOT SUB: Drag Reduction by Use of Heated Boundary Layer
  —Which Airfoil Design Will Produce the Most Lift?
☐ Human Psychology/Social Science
  —Racial Stereotypes: What do 2nd and 3rd Graders Believe?
  —Correlation Between Left/Right Brain Hemisphere Dominance and Inkblot Responses

☐ Mathematics
  —Analysis of a Championship Season—the 1984 Padres
  —Randomness in the Game of Life
☐ Medical Science
  —Stress: The Key Element in the Development of Diabetes
  —Self-Relaxation vs. REM Sleep to Lower Blood Pressure
☐ Physics
  —Influence of Textured Surfaces on Magnus Effect Rotors
  —Looping Roller Coasters—Can They Be Safe?
☐ Zoology
  —Effects of Shaving on Tarantula Thermotropism
  —Do Pesticides Affect Earthworms?

Categories, as defined by the International Science and Engineering Fair, are somewhat broader. A list of these, together with their descriptions and interpretations, appears in Appendix A.

FINDING A TOPIC

Besides finding a topic that you like, be sure to select

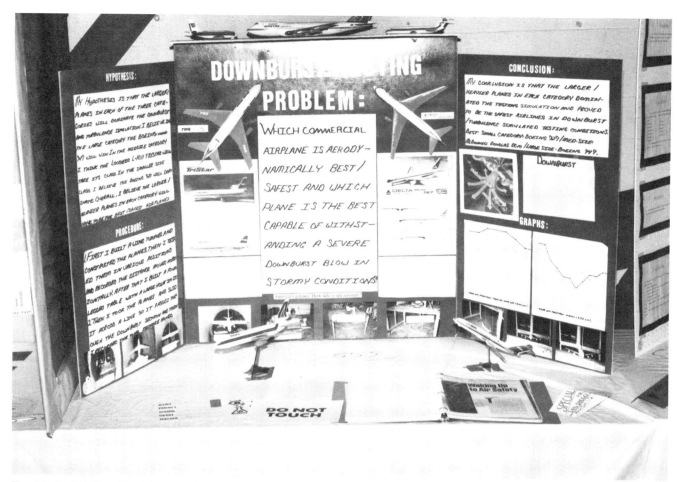

Fig. 2-2. Brett Callan; *Which Commercial Plane is Most Stable in Downburst/Turbulence Conditions?*

a subject with which you feel comfortable and confident. Although the main purpose of doing a science project is to explore new techniques and learn new things, it is best to pick a topic in which you have at least a basic understanding. This is not a good time to do a crash self-study course in your most difficult subject. For example, if mathematics are giving you trouble, a topic requiring complex and sophisticated calculations is unwise.

If this is your first project, or you have not yet taken many science courses, you may be wondering how to find a good idea.

Of course, your math, science, or computer classes are excellent sources. Even classes in social science might suggest an excellent topic idea. For example, a class discussion on subliminal advertising led to an Animal Behavior experiment that measured the effect of sound on mouse appetite. History courses that deal with prehistory might lead to a project dealing with fossils, or dating ancient artifacts.

You might also find a question that arouses your curiosity in a newspaper or magazine. Areas of science, especially those that are important in your particular locality, are often discussed in feature articles. Weekly news magazines deal with the areas of health, aerospace, the

environment, and other scientific fields, particularly when new problems, discoveries, or theories are presented.

Problems affecting teenagers are often the inspiration behind projects. The attention on drug-related problems no doubt inspired Doug S. Fay's project, *Drugs Weave Patterns,* which indicates the degree to which alcohol, caffeine, and nicotine impair the spider's ability to weave its web. Catherine Jacobus' project, *The Effects of Selected Carbohydrates on Alcohol Absorption and Behavioral*

Fig. 2-3. Illustration of light bulb.

Modification in Humans, which attempted to find a way to block alcohol's effects, also relates to a very timely issue.

The problem of world hunger inspired Neil Ireland's experiment, which measured the effect of seed size on wheat. Projects on acid rain and pollution levels are also popular choices for many students, for example Leslie S. Thomas' project *Acid Rain—Causes, Effects and Cures, A Seven Year Study,* in which he rehabilitated two ponds that had been previously acidified, or Eric Felt's project, which tested how safe the EPA's standard acceptable levels of toxic wastes and herbicides really were.

Movies or television shows, even if they are fiction, may deal with scientific subjects. For example, several years ago, a successful psychology project used a survey concerning children's reactions to the television film, *The Day After.* More frequent, (perhaps almost to the point of being overdone) are projects dealing with the effects of Rock music (with or without suggestive lyrics), MTV, or video games on heartrate, grades, or a variety of other measurable factors.

A problem, such as a shortage, a natural disaster, an unusual weather condition, or a newly discovered pollutant or health problem, might spark your interest. For example, the energy shortage several years ago inspired J. J. Tessier's project comparing different types of building insulation.

Sometimes, even if you cannot research the condition itself, you might build a project around people's reactions to it, or the psychological impact of a certain issue.

Instances of current, local problems inspiring a science project are the study of the effects of the Tijuana River pollution on the San Diego county beaches, which has become a political as well as an environmental issue in that part of the country. Susan Strickland's experiment, inspired by a local environmental problem rated the flame retardance of various plants that could be grown in the Southern California canyons.

Not only local weather conditions, but the local geography, weather and wildlife itself is often the inspiration for science projects. Living near the ocean usually stimulates interest in the sea itself, as in the Kerry Kostman's project *The Effect of Weather Patterns on Sea Water,* or Ted Larson's marine biology study, *A Comparison of Ocean and Bay Plankton.*

So, the best way to find a topic is to be aware and receptive to your environment. The world around you offers many opportunities for topics.

In talking with science fair participants, we found a variety of reasons for choosing a particular topic.

Fig. 2-4. J.J. Tessier; *Effectiveness of Rigid vs. Batt Insulation.*

Fig. 2-5. Susan Strickland; *Fire Danger from Native Shrubs.*

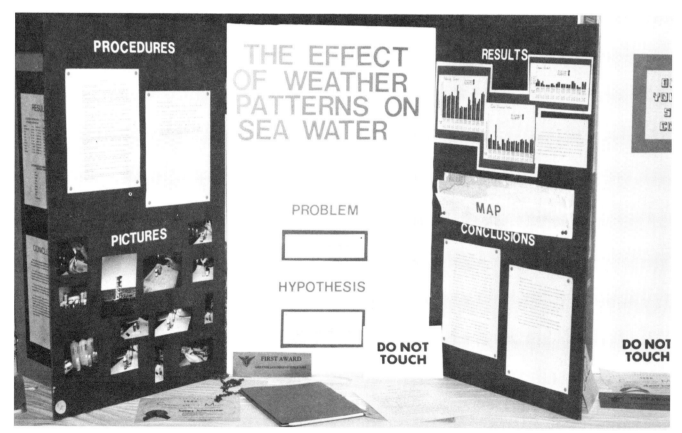

Fig. 2-6. Kerry Kostman; *The Effect of Weather Patterns on Sea Water Composition.*

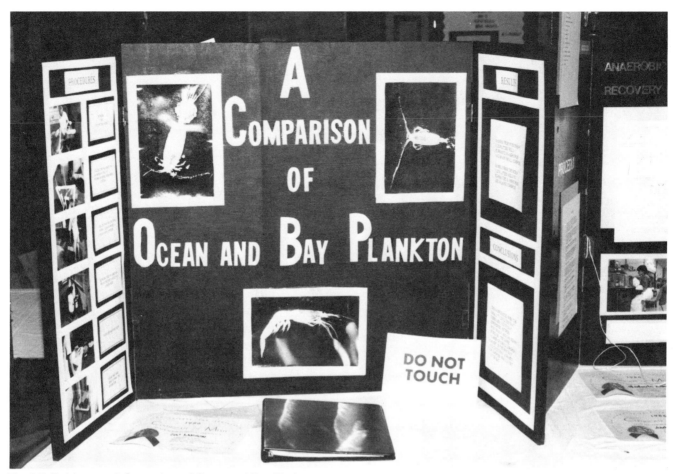

Fig. 2-7. Ted Larson; *A Comparison of Ocean and Bay Plankton.*

"I wanted a relatively simple project with interpretable and quick results," said Nicola Kean of Redlands, CA, discussing her project *What Happens When the Length of a Pendulum is Increased?*

Sometimes, an idea is suggested by observing a particular phenomenon. Maryanne Large of Sydney, Australia, had been unsuccessful in finding a topic when she accidentally observed an effect while printing some photographs at home. From that, she developed her project topic, *Visual Reversal of Photographic Negatives.*

Other concepts resulted from a long standing interest in a particular field. Carl T. Donath of Lafayette, New York, was interested in the idea of a voice or speech recognition system for a computer, and consequently developed a system for *Computerized Speech Analysis and Recognition.* In attempting a project of such scope, Carl had the help of his father, an electrical engineer.

Keith Maggert of San Diego, California is another participant who was inspired by his father's work. When the government denied his father a grant to conduct research to determine the importance of hair to a tarantula, Keith decided to find out. For two years he conducted experiments dealing with the body heat of tarantulas.

Another reason for a choice of topic is to fulfill some

particular objective. Ben Cheng, with his project *The Role of Amoebocytes in Zinc Uptake and Transport in the Snail Helix Asperia* wanted a project that would contain factors important to the training of a modern experimental biologist. Although the project was included in the Zoology category, it combined both cellular biology and environmental science.

Some of the participants were quite frank and open about their reasons for choosing a particular subject. "Notoriety and recognition" Billy H. Scott of Jacksonville, Florida told us, when asked why he selected his project *Obtaining Kinetic Energy through Heat,* using rubber bands to 'build' his machine. "I wanted something outstanding, which would attract attention."

A word of warning. Well meaning friends and relatives may propose or try to impose ideas, and even offer their assistance. Before you make a decision, however, *be sure* that *you* are interested in the topic, and have the knowledge and skills to carry it through to completion *on time.* A project that relies solely on the knowledge and work of parents, friends, and relatives will not help you to achieve any of your goals in doing a project. If you are uncertain, discuss the idea with your teacher or advisor. He or she will be glad to talk with you and may even point out something (either

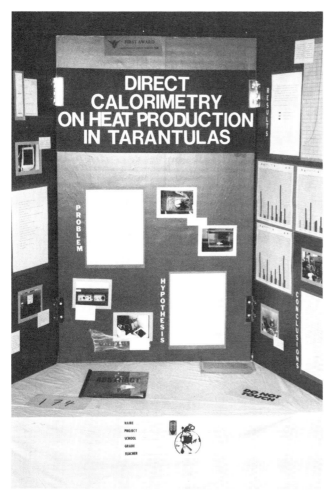

Fig. 2-8. Keith Maggert; *Direct Calorimetry on Heat Production in Tarantulas.*

positive or negative) that you hadn't yet thought of.

CAN I DO IT?

Once you find a topic (or have several possibilities) it's a good idea to analyze the feasibility of your plans. By this, I mean that you must now decide whether you can actually complete the project in the time available. Here's a checklist to help you decide:

☐ Is the information that I need readily available to me?
☐ Where can I find the information?
☐ If the information is not available locally, where is it? How long will it take to get?
☐ Will I need to pay for my information? (government agencies or industry may charge for their data).
☐ If I need special books, can I check them out of the library or must I use them there?
☐ Will I need professional advice? From whom? (specifically!!)
☐ Are they willing to help? Will it cost anything?
☐ What supplies will I need?
☐ Can I build some of the things I need? Do I need help?

☐ Can I borrow some of the supplies and equipment I'll need?
☐ What will my supplies cost?
☐ Do I have the money for my supplies and equipment?
☐ Will my parents let me leave home with their Gold Card?
☐ Can I finish in the time allowed? If not, can I break the project idea into smaller segments?
☐ Is there anything about the experiment my family might object to?

Answering these basic questions will, at least, give you a rough idea of whether you can complete this project. I repeat (at the risk of being boring), if you have any doubts, consult your teacher or advisor. However, if you are sure that this is the topic for you, now is the time to consider some of the elements of your experiment in more detail.

SAFETY FIRST

If you are dealing with chemicals or electricity, you must insure your personal safety as well as the safety of your environment. Some questions that should be answered are:
    Electricity:
☐ Is battery power adequate?

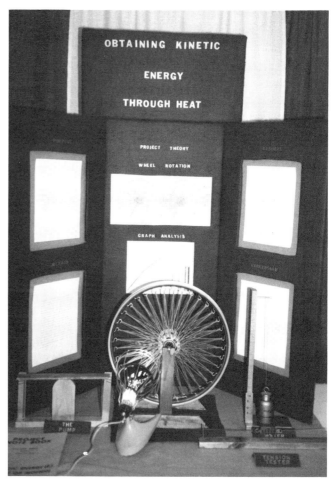

Fig. 2-9. Billy Scott; *Obtaining Kinetic Energy Through Heat.*

- ☐ If I plan to use house current, is there sufficient wiring?
- ☐ Can I avoid overloading circuits?
- ☐ Will everything be properly insulated and grounded?
- ☐ If I must use line voltage, do I need an electrician?
- ☐ Do I need the approval of an electrical inspector?
- ☐ Do I need to notify the local power company?
- ☐ Is the power the right kind (110 vs. 220 vs. 440) or do I need transformers or converters? Are they readily available?
- ☐ Is there any danger that I could blackout seven states and the island of Guam?

Chemicals:
- ☐ Are the substances readily available, or do I need special permits to acquire them?
- ☐ Is anything I plan to use or synthesize considered a toxic substance?
- ☐ Do I need approval of state, local or federal authorities (such as the EPA) to do my experiment?
- ☐ Are there any unstable compounds which could possibly cause an explosion? Will these compounds, either alone or in combination with other substances form toxic gasses, or start fires that might cause injury or property damage?

- ☐ Will I need special containers?
- ☐ Do I need protective gear (goggles, aprons, gloves) to work with this substance?

Very possibly, you may need adult assistance or supervision to insure a safe experiment and environment with an electrical or chemical project.

LIVING LABORATORY

Many experiments in the animal behavior and zoology categories rely on the use of live specimens. Although these types of topics may provide some of the most fascinating and fun projects (imagine the thrill of teaching a gorilla to talk!), this introduces another factor into your science project plan.

Carefully consider whether you can and will commit the time and effort to care for these animals. Fortunately for you and your family, the project probably won't involve anything as large or exotic as an ape, but even the smallest live vertebrate requires special handling. Another factor in deciding whether you want to do an experiment of this type is to determine exactly how many specimens you need. You may need a rather large number, kept in separate groups,

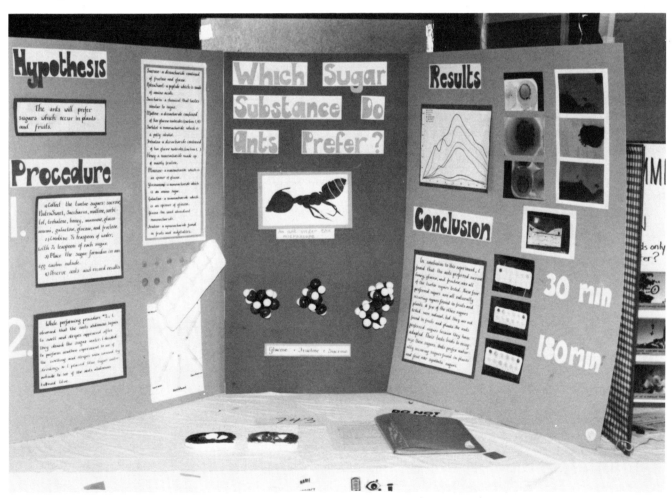

Fig. 2-10. Tasha Taylor; *Which Sugar Substances Do Ants Prefer?*

Fig. 2-11. Billy Scott; *Obtaining Kinetic Energy Through Heat* (close-up).

for your experiment to be scientifically valid. Be sure you have the facilities to keep three cages of twenty mice each completely segregated, before beginning this type of experiment.

To do this properly, learn all you can about the animals' care, feeding, habits, and temperature needs, as well as their cleanliness requirements. You must also research the reproductive cycles and characteristics of the species in order to plan for possible periods of gestation and birth during the course of the experiment. For example, when dealing with mice or other rodents, it is extremely important to remove the males when a litter is born, to prevent them from eating the newborns! Many students using rodents have therefore limited their specimens to one sex (provided, of course, that reproduction is not a factor to be considered).

It is also important to understand that some types of laboratory animals are cannibalistic, particularly when they are kept in close quarters. Be sure you can deal with this fact of animal life. If you find it upsetting, choose another topic.

Another word to the wise—be sure to get your parents' approval before embarking on such an experiment. Nothing will terminate your project faster than an unknowing parent entering the house to find 600 (or even 60!) mice running around. Talk to Mom and Dad. You will need their cooperation. If you do get their approval, be conscientious

in keeping the animals and their quarters clean. Help and cooperation will fast disappear in a home that smells like a neglected cage in a travelling zoo!

When working with live vertebrate specimens, be aware of the role of animal rights lobbies. To insure that animals are not mistreated, you must have an expert certify that you are capable of caring for the animals and that your experiment will not harm them in any way. Judges may want explanations of any deaths in your experimental or control groups. Irregularities can prevent you from exhibiting at a science fair.

We have also found, in talking with exhibitors and coordinators from all over the country, that some local fairs use different standards for judging experiments using live specimens. To be sure that your project will be acceptable if you move ahead in competition, adhere to the standards required by Science Service. They conduct the International Science and Engineering Fair. By doing this, you will be able to compete anywhere. You can get a copy of the ISEF Rules by sending $ .50 to:

Science Service
1718 N Street N.W.
Washington, DC 20036

If it turns out that caring for live vertebrates is not for

you, this doesn't exclude you from working in certain other categories. Try insects. Tasha Taylor won a First Place with an Animal Behavior project entitled *Which Sugar Substance do Ants Prefer?*

On the other hand, you can use vertebrates in a project, while not needing to experiment on them for long periods of time. In *The Canine Allergy-Obesity Connection* by Missi J. Wildenfield, most of the work consisted of recording the dogs' body weight and tabulating these figures together with criteria established for allergic conditions. The only "live" tests on the dogs were pinch tests to determine the body fat content.

DOLLARS AND CENTS

Before making a final decision, estimate all your total costs. It is not necessary to spend a huge amount of money to be successful, nor must you have access to a sophisticated million dollar computer to do a thorough and creative experiment. Some very successful projects, which have advanced all the way to international competition, have been done with minimum expenditure, whereas others require a great deal of money. Students reported that their projects cost anywhere from $12.00 up to $400. Billy Scott wound up spending $150, mostly for light bulbs.

Often, laboratory equipment, as well as computers and other machines, can be borrowed or used on high school and university campuses when classes are not in session. If you think you will need facilities of this type, arrange for it before you commit too heavily to your idea. Stephen M. Jacobs told us he spent approximately $375 for equipment and supplies, but including borrowed equipment, the value of the project was about $13,000.

Caroline K. Horton, for her project on Fly Ash Foam, borrowed expensive equipment from companies and plants. "This," she said, "makes it possible to carry on research that would otherwise be impossible due to the cost factors."

Try to be as accurate and complete as possible when making your estimate. For many exhibitors we interviewed, the project ended up costing more than they had anticipated. The biggest exception to this was where the experiment was a continuation of a prior year's project, it was easier to budget the expenses. When you're making your estimate, include everything, from the supplies for the experiment to the materials for the graphs and the display board. Often, people forget that art supplies are sometimes the most expensive portion of the project.

In any event, making an estimate, and perhaps even budgeting or saving for these expenses, will help you avoid unpleasant surprises at the last minute.

One last step is to get the approval of your science teacher or advisor. Sometimes, you'll need him or her to sign a form, agreeing to the experiment.

Now that you've selected your topic, let's begin!

Fig. 2-12. Stephen Jacobs; *Design, Construction and Operation of a Continuous/Pulsed Ion Beam Accelerator* (close-up).

STUDENT PROPOSAL FOR SCIENCE PROJECT
MUST BE COMPLETED AND APPROVED BY TEACHER-ADVISOR BEFORE START OF EXPERIMENTATION

TO: _____ TEACHER

_____ HIGH SCHOOL

STATEMENT OF PROBLEM:

PROPOSED EXPERIMENTAL PROCEDURE: (Use back of sheet if necessary)

LOCATION WHERE EXPERIMENTAL PROCEDURES WILL BE DONE: _____

PROPOSED PROJECT CATEGORY: _____

I.   **DOES EXPERIMENT INVOLVE LIVE, VERTEBRATE (NON-HUMAN) ANIMALS?** ☐ YES ☐ NO

IF YES — THE QUALIFIED ADULT WHO WILL SUPERVISE THE EXPERIMENT IN ACCORDANCE WITH THE REQUIREMENTS OF FORM GSDSEF-2 — "CERTIFICATION OF HUMANE TREATMENT OF LIVE VERTEBRATE ANIMALS"* IS:

NAME _____ _____ DEGREE: _____ TITLE: _____

ADDRESS: _____ PHONE NUMBER _____

II. **DOES EXPERIMENT INVOLVE HUMAN SUBJECTS?** ☐ YES ☐ NO

IF YES — TO ENSURE THAT NO PHYSICAL, PSYCHOLOGICAL OR SOCIAL RISK IS INVOLVED, CAREFUL PLANNING WITH MY TEACHER-ADVISOR WILL INCLUDE STUDY AND COMPLETION OF FORM GSDSEF-3 — "CERTIFICATION OF COMPLIANCE OF RESEARCH INVOLVING HUMAN SUBJECTS."*

III. **DOES EXPERIMENT INVOLVE TISSUE SAMPLES OF HUMANS OR VEREBRATE ANIMALS?** ☐ YES ☐ NO

IF YES — THE INSTITUTION OR BIOMEDICAL SCIENTIST WHO WILL PROVIDE THE TISSUE IN ACCORDANCE WITH THE REQUIREMENTS OF FORM GSDSEF-4 — "CERTIFICATION OF TISSUE SAMPLE SOURCE"* — IS:

Name _____

Address _____ PHONE NUMBER _____

IV. **DOES EXPERIMENT INVOLVE RECOMBINANT DNA TECHNOLOGY?** ☐ YES ☐ NO

IF THE ANSWER TO EITHER I, II, OR III ABOVE IS **YES, I AGREE TO STUDY** AND **COMPLETE** THE NECESSARY CERTIFICATION, FORM GSDSEF-2 GSDSEF-3 or GSDSEF-4 **BEFORE** BEGINNING MY PROJECT.

IF THE ANSWER TO IV ABOVE IS YES, I WILL CONTACT ONE OF THE SCIENCE FAIR OFFICIALS LISTED IN THE STUDENT GUIDE TO ENSURE COMPLIANCE WITH REGULATIONS AND COMPLETION OF NECESSARY FORMS RELATED TO RECOMBINANT DNA RESEARCH.

_____   _____
            Student Signature   Date

IN AGREEING TO SPONSOR THE STUDENT NAMED ABOVE, I CERTIFY THAT I HAVE REVIEWED THE PROPOSAL, THAT IT COMPLIES WITH THE RULES OF THE GREATER SAN DIEGO SCIENCE AND ENGINEERING FAIR, AND THAT, IF APPLICABLE, FORM GSDSEF-2, GSDSEF-3 or GSDSEF-4 HAS BEEN COMPLETED AND/OR THE STUDENT HAS COMPLIED WITH ALL THE ABOVE REQUIREMENTS.

_____   _____
         Teacher-Advisor Signature   Date

*Form available from science teacher

Form GSDSEF-1 (1987)

Fig. 2-13. Student proposal for science project.

Chapter 3

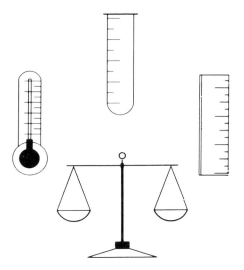

Getting Organized

Now that you've chosen your topic, you are ready to begin organizing your project. Since this is probably the largest assignment you've ever done, the planning phase is the key to your success. Because of the length of time involved and the variety of activities you'll be conducting during the course of your science project, organization is crucial to this venture.

One of the biggest pitfalls in a long-term assignment is the tendency to leave things until the last minute (Take heart—this tendency to procrastinate is not limited to junior high and high school students—many people battle with this problem through their college years and into their careers!) It is, after all, easy to take advantage of a warm October day to ride your bike or play football instead of going to the library when the project isn't due until February!

If this project is required, your teacher has probably provided intermediate deadlines for you. This will help you to complete your project on time by giving you short-term goals. Even experienced students and many adults need milestones, with concrete results and items due, in order to keep going on a large project.

A schedule with short-term due dates will also help your teacher assist you, giving him or her the chance to help you correct small problems before they become insurmountable hurdles. Such a schedule may also help you identify problem areas, giving you the chance to change course while there is still time to find other resources.

If this project is not required, you may have to set up your own deadlines and schedules. Below is a "time line," one way of presenting a science project plan and schedule of due dates.

Because sheets like this tend to get lost, however, the schedule might be more useful if you transfer it to a large wall calendar and put it up in your room, where it will be a constant reminder.

You will notice "Log Check" occurring several times in this schedule. This is not a check of your Experimental Log, which is a diary of your experiment, and which we describe in Chapter 6. This Project log is a record of the entire project itself. Its purpose is to keep you organized and "on track." It is an informal record of all your activities on the project. This should include:

a. Research
b. Consulting the experts
c. Finding and buying supplies
d. Building equipment, if necessary
e. Setting up the experiment
f. Performing the experiment including testing and observation
g. Data analysis
h. Typing the paper
i. Building the backboard

The best way to keep a time log is to draw up a form, so that you can simply "fill in the blanks" as you work. Keep it available, so that you can make an entry every time you work on the project.

Even if this is not required, good records may prove useful later on. For example, one participant who needed data on solar activity contacted a division of the government agency, NOAA, who referred him to several U.S. and private agencies. One of these sources turned out to be a research institute library right in his own city! However, having the entire list might have been useful in case that resource did not turn out to be productive.

Documenting your activities could also be useful if you

Timeline—Due Dates of Substeps	
Date	*Item*
9/25	Topic Card, in the form of a question
10/4	Log Check
10/10	Bibliography Cards (at least 7 references)
10/20	Log Check
11/1	Notecards (at least 100, cross-referenced by source)
11/15	Log Check
11/20	First Draft - background research paper
11/30	Hypothesis, Procedures & materials list
12/1	Log Check
12/10	Revised hypothesis, Procedures & materials list (if necessary)
12/15	Background Research Paper (final)
12/20	Log Check
1/8	Log Check
1/17	Log Check
1/25	Log Check
2/5	Results (tables, graphs, etc.) and Conclusions
2/10	Project Notebook
2/20	Project Display
2/28	School Science Fair
3/31	City/County Science Fair

Fig. 3-1. Timeline schedule.

Table 3-1. Project Log.

Name_____

Project Title_____

Time Log

Date	Time	Activity
10/25/84	3:35 P.M.	Visited central library to find source material on eating habits of mice.
11/04/84	4:30 P.M.	Spoke to J. Jones, owner of Pet-a-rama pet shop to find out if 60 mice are available and how much the animals, cages and food would cost.
11/07/84	10:00 A.M. thru 2:00 P.M.	Spent time in library doing research for background research paper.

do not receive the responses you requested, or if delivery is late or non-existent on materials you ordered. This means not only making an entry in your log whenever you call or speak to anyone, but also keeping copies of any letters you send or receive in the course of your project. It will be easier to explain why you missed your deadline when you have documented evidence that you really did order your fruit flies back on November 1st, and called to inquire every Tuesday since that time!

Once you've drawn up a plan and a schedule, make sure that you let it be a tool to help you manage the project and not allow it to be something that rules you. If you find that the form you developed doesn't suit your needs, for example,

not allowing enough room to write all your notes, *change it!* Do not allow small annoyances to become large frustrations that prevent you from keeping accurate records.

Somewhere during the course of the project, if things aren't going the way you planned, show your teacher your records and explain the reasons you think the plan needs to be changed. He or she will be glad to help you work out the problem and advise you, but the task will be much easier if your documentation is good.

Remember that the purpose of organizing the project is to help you be successful, and not to restrict you to plans or procedures that are not working.

Chapter 4

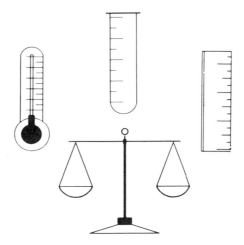

Writing the
Background Research Paper

A background research paper is one of the required parts of a science project. For some of you, writing a paper of any kind is a more fearsome prospect than doing the experiment. You may imagine entire weekends spent unearthing dull facts. However, by breaking the process down into manageable steps, it can be less intimidating and more fun.

WHY DO A PAPER?

Before you begin, it will help to understand why the paper is necessary. There are three reasons that a research paper is required.

In order to conduct any worthwhile experimentation or research, it is essential to gather as much information as possible. This will help you to understand the theories, research and discoveries that have already taken place. It should provide a great deal of insight into your topic, and also save unnecessary steps in trying to prove some already established facts. Your research can also help you to narrow down your topic, particularly if the subject you've chosen is very general. This will be helpful later on, when you're trying to formulate your question and hypothesis.

Another reason applies to your careers as students. Throughout your education, countless papers of various lengths will be required in all subject areas. Some of these papers may be pure research, calling for gathering and organizing factual data. Others will be of a more interpretive nature, where you will need to learn the facts, and then show your understanding or opinion based on the knowledge you've gained. In any event, it's vital to know how to plan, organize and write the research paper.

This requirement also acquaints students with the actual process of conducting research. The ability to track down information is a skill you will use throughout your lives, even when your careers as students are over. As working adults, parents, or simply curious or informed citizens, the ability to find facts is essential. If your parents are helping you with this research, they'll know what I mean!

Actually, once you get started, this phase of your project can be lots of fun. In many ways, it's almost like playing detective. You may find one source that is not especially useful, but can offer a clue that may lead to three or four new sources. As with any other mystery, your research may lead you down some blind alleys, which will temporarily throw you off track. In the end, though, if you persevere, you'll certainly find what you are looking for. Along the way, you can learn many more fascinating things, perhaps on other subjects leading to new interests.

FINDING THE FACTS: A BRIEF GUIDE

The first step in the process is locating your information. A natural place to start is with the sources that are closest to you. These include your textbooks, any encyclopedias or other reference materials you may have at home, and publications in your school library. These references may yield only the most basic information, but will at least help you to focus on the various aspects of your topic. Also, the bibliographies in these volumes might point out other sources.

The next stop is your public library. If you do not know how to use the catalog there, ask the librarian's help. Local branch libraries often use card catalogs, employing the Dewey decimal system.

Other libraries use the LC, or Library of Congress, method to catalog books.

Both methods will guide you to the areas of the library

Science Fair: Developing a Successful and Fun Project

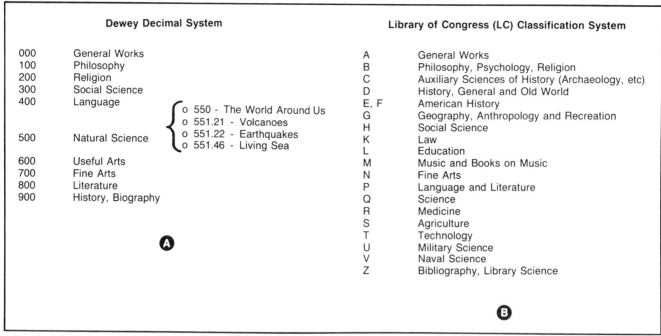

Fig. 4-1. Dewey decimal system (A) LC system (B).

where you'll find materials dealing with your subject.

Some libraries use a microfilm catalog in addition to the card catalog. This usually shows everything available in the entire library system. If the resources you need are in other branches, you may be able to order them delivered to your branch library. Otherwise, you'll need to travel to other parts of your city or county. Eventually, your search will probably lead you to your library's main branch to get all the information you need. If you live in a large city, this may be the opportunity to get acquainted with a sophisticated, computerized catalog system. This type of search will locate materials based on either author or subject.

Don't get the impression that you have to live in a big city to do research. To begin, of course, you'll have your school and local libraries. Then, all you need is writing materials and a post office, and you can gather material from all over the world. As you'll see later in this chapter, publications dealing with just about every conceivable area of science are available from the U.S. Government. Private companies are also cooperative, and quite willing to share information. Even foreign governments and international organizations might be willing to send material. The whole world's available for the price of a postage stamp!

For the moment, however, back to the library. Some materials may be available for you to check out. Before you bring the book home, scan it briefly to see that it really meets your needs. Check the copyright date to see if it's sufficiently recent to contain the most up-to-date information on your subject. For example, a book on nuclear energy written prior to 1950 would hardly be worth your time! In some rapidly expanding scientific fields, even information that is a few months old can be obsolete. Also, be sure that the material

is on your level. A manuscript written for use by a Ph.D. in biochemistry may be far beyond your capabilities if you're a seventh grader attempting your first science project. Looking at the table of contents, index and appendices might also give you a good idea of whether this book will be useful for you.

Other materials will be classified as reference material, which means you will have to use them at the library. If you need to do that, please remember library etiquette. Keep your materials close at hand and neat, talk in whispers, and no eating or drinking. Be sure you have plenty of paper and pencils, because there's usually no place to buy anything at the library.

Another way of doing research is to make copies of important information to bring home. Most libraries today have photocopiers, which are most helpful. But be prepared and bring plenty of change. These machines gobble up money faster than a video arcade!

Do not limit your research to books. There is a lot of valuable information in periodicals. This includes popular magazines, such as *Newsweek, Time,* and *Readers Digest,* and magazines that popularize scientific subjects, such as *National Geographic, Today's Health,* and *Psychology Today.* There are also specialized, scholarly periodicals that may deal with your subject. To find these types of articles, use a periodical index, such as *The Readers Guide to Periodic Literature.* Some of these indexes are:

—*Abstracts of Popular Culture*
—*Access*
—*Index to Free Periodicals*
—*Magazine Index*

18

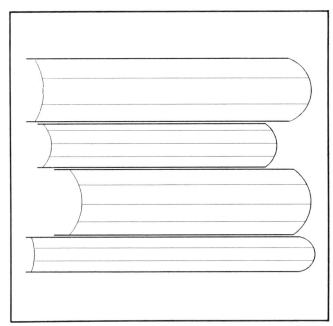

Fig. 4-2. Illustration of stack of books.

One of the most extensive sources of information is the U.S. Government, our nation's largest publisher. In fact, the various government agencies publish such a large variety of materials on a vast range of subjects, that in Lois Horowitz' book, *WHERE TO LOOK, the Ultimate Guide to Research,* the chapter on Government documents is the longest!

One way to locate helpful government references is to look at the list of U.S. government departments and agencies that publish extensively. Here is a list of just a few that publish scientific documents:

1. Department of Agriculture
2. Department of Commerce
3. Department of Defense
4. Department of Education
5. Department of Energy
6. Department of Health and Human Services
7. Department of the Interior
8. Department of State
9. Department of the Treasury
10. National Aeronautic and Space Administration
11. National Oceanographic and Atmospheric Administration
12. Bureau of Mines
13. Fish and Wildlife Service
14. Public Health Service
15. Atomic Energy Commission

Fig. 4-3. List of government sources.

Once you have located your sources, you'll find that if you require a recent edition, the library will usually have the actual issue on hand. Older issues will probably be available on microfilm or microfiche. Occasionally, the microfiche reader will have a copier attached (which also takes nickels, dimes, and quarters), so that you can make copies of important pages.

College and university libraries can also provide an invaluable resource for you. Often, they will have books and periodicals that a public library does not carry, especially if the college is noted for a particular area of science. The people there, who are sometimes student workers, can be very helpful in guiding you to the facilities available.

If a conveniently located university library has a great deal of useful information for you, investigate getting a library card. Many universities issue these cards for a fee, which is sometimes a tax deductible contribution. This will allow you the same library privileges as the college or university students, enabling you to check out nonreference materials.

While you're there, see if you can buy a copier card. This works something like a bus pass, because you prepay the cost of making a specific number of copies on the library's machine. This usually offers a saving over the cost of individual copies. It's also quite handy, because you don't have to worry about carrying a pocket full of change.

Institutes and foundations usually function the same way as university libraries. Without special access, their materials are only available for use on the premises, and cannot be taken out of the building (there are copiers here too)! However, these can be extremely good sources of specialized information. The information you find in institute libraries can perhaps lead you to experts in your field.

In your search for relevant government material, don't only look in the obvious places. The Department of Commerce, for example, publishes an annual entitled *United States Earthquakes,* and the Department of the Treasury, which is in charge of Alcohol and Tobacco, has many publications on those subjects.

There are several indexes to help you wade through the large amounts of government material. Several of these are:

1. *Monthly Catalog*
2. *Cumulative Subject Index*
3. *Publications Reference File*
4. *Index to U.S. Government Periodicals*

Even small libraries should have access to at least one of these research tools. The librarians may be able to help you use them. Once you have located the reference, it may not be readily available to you. Note, however, that some government publications can be purchased from the appropriate agency. Be sure you allow sufficient time to receive the materials you order. Many a project has faltered waiting for an order to be processed.

Branches of the military service, corporations and professional associations are also excellent sources of information on a variety of scientific and engineering topics. Often, they publish pamphlets which are yours for the asking!

Computer Note:

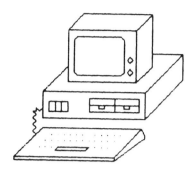

If you have access to a large computer data base, such as *CompuServe* or *The Source*, it can help focus your research by providing a list of resources. However, be sure to plan your searches before you begin. Some of these time-sharing services charge different rates depending on the day of the week and time of day, much like using the telephone long distance. (Incidentally, don't forget about your phone charges when estimating the cost of using a large data base.) Also, because you're charged by the length of time you're connected, consider having the print-out mailed to you rather than checking the list on your screen.

As you look for your sources of information, review this checklist to be sure you've left no stone unturned!

Table 4-1. Resource Checklist.

Places	Things
_____ School Library	_____ Encyclopedia
_____ Public Library	_____ Books
_____ University Library	_____ Magazines
_____ Institutes	_____ Scholarly journals
_____ Industry	_____ Professional journals
_____ Government	_____ Newspapers
_____ Military	_____ Government publications
_____ Zoos, Seaquariums	_____ Data Bases
_____ Museums	_____ Local newspaper

TAKING NOTES

Now that you have located the information you need, you'll need to extract the most meaningful data to use in your background research paper. Because every fact you use will be derived from one of your sources, the first task is to start a working bibliography. Although this is a rather informal list, without regard for the "official" bibliography format you will use when submitting your final paper, include all the vital information about your source. Code each source with a letter or number, which you'll use later when writing up the index cards.

Examples of items in a working bibliography are as follows:

> A. Encyclopedia International, Vol 6, Earthquakes
> B. "Is California Sinking?," Scary Science Magazine, 1984

Fig. 4-4. Working bibliography.

You will, of course, want to take notes efficiently and effectively. The key to accomplishing this is to follow two important (and entirely contradictory) rules. First, focus on the most important and relevant facts, analyzing every piece of information, to determine whether or not it truly relates to your topic. Secondly, keep an open mind. Sometimes, a new, seemingly irrelevant fact can shed new light on your subject.

You will often find that information on several similar topics is included in one article or book. Extract only the information that furthers your understanding of the subject. If you find the same facts in several sources, use the presentation that makes the most sense to you.

Write each fact, in your own words, on an index card. It is important to use *your own words*. This will help you understand and absorb the material. However, if it's important to quote the material word-for-word, be sure to note that it's a direct quotation. Code each card to show the source of the fact and the page number.

Limit each index card to one fact only. This may seem like tedious work as well as a waste of space, but this method will make it a lot easier for you to write the final research paper. In addition, your science teacher may require you to turn in a specified number of index cards, along with your bibliography, as evidence of your research.

Computer Note:

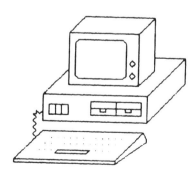

If you have access to a computer, you may wish to use your word processor or data base software to create your index cards. In this way, your data may not have to be recopied when it's time to create your final research paper. Index cards are available for any tractor feed printer. Consult your software user manual to set up the print format for 3 × 5 cards.

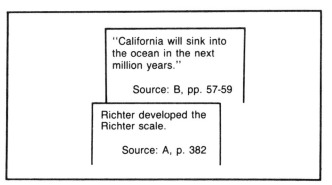

Fig. 4-5. Index cards.

OUTLINING

Before you begin the actual writing, take the time to review all your facts. Try to determine a few themes, or main topics, which tie the material together. As you do this, some subtopics should also become apparent.

Write down, in outline form, what you want to say in your background research paper. If you are not well acquainted with using outline form, it is:

```
    I. Main topic I.

        A. Subtopic A
        B. Subtopic B
            1. Undersubtopic 1
            2. Undersubtopic 2
                a. sub-under-subtopic a
                b. sub-under-subtopic b
                    i. under-sub-under-sub-etc. i
                    ii. under-sub-under-sub-etc. ii
    II. Main topic II.
```

Fig. 4-6. Outline form.

Remember that in outline form, if an item is subdivided, it should contain *at least* two elements. Otherwise, the information should have been included in the division above. For example:

> IV. Other insects
> 　　A. Tarantulas

should read:

> IV. Other insects; tarantulas, etc.

Once you've formulated the outline, you'll see how easy it is to arrange your facts. Use the outline to divide your index cards into stacks that relate to each topic and subtopic.

Don't worry if you have a few cards left over that do not seem to belong anywhere. They may either represent facts you've gathered that are irrelevant to your subject, or may point out a need to include another topic or subtopic in your outline. One helpful hint—you may want to keep these extra cards although you don't want to include the facts in your background research paper. Data you may not need now could be extremely important as you progress in your experiment. Also, if you think towards the future, some science project participants build on their project year after year, going on to some new, original aspect each time. These facts that do not fit this year's experiment might be your guiding inspiration next year.

Next, arrange each stack of cards into a logical sequence. These ordered stacks of cards will now become the basis of the first draft of your research paper.

WRITING THE PAPER

Although your facts are now organized, let's take a moment to review the format of a well-constructed research paper. The first few paragraphs should introduce your paper, stating the topic you're writing about and the experiment you think you will do. Then, follow your outline to provide a logical presentation of your background information. The last several paragraphs should summarize the facts you've discussed, and finally, you should predict what you believe your experiment will prove or disprove.

Computer Note: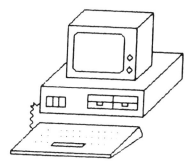

If you have recorded your facts using a data base or a word processor, use your software to rearrange your facts in the same sequence as you've classified your index cards. Delete any unrelated facts and store your file under a *new name.* This will preserve a copy of your old list "just in case."

If you've recorded all the facts you've found in your own words, it will be easier to create a research paper that is truly your own. Remember, however, that the paper does not simply consist of a bunch of facts strung together. Your teachers, as well as science fair judges, will look for some creative thought and educated interpretation of the facts, which tie your research to the experiment you will conduct.

One resource you'll find valuable as you begin to write is Strunk & White's *Elements of Style.* Originally published in 1919, it is the definitive guide to good, clear writing. This small, inexpensive paperback is one you will use frequently throughout your student career. The book not only describes rules of correct grammar, but also gives guidelines for writing clearly and concisely. Especially for inexperienced

---

1. For the first footnoted reference to a book:
   [1]Robert W. McLuggage, *A History of the Americal Dental Association* (Chicago, American Dental Association, 1959), p. 475
2. For subsequent footnoted references:
   [4]McLuggage, pp 15-18
3. For a periodical
   [12]*Psychology Today,* October 1986, 68

(NOTE: In all cases, the materials cited in your footnotes should appear in your bibliography, in proper format, as well.)

Fig. 4-7. Footnote format.

writers, the examples of misused, overworked, and common, incorrect expressions might be the book's most helpful feature. Following even a fraction of the helpful hints in *Elements of Style* will help you create a more polished and readable research paper.

CREDIT YOUR SOURCES

At this point, we must discuss an unpleasant little topic, called plagiarism. It's a topic you'll hear discussed throughout your educational career. Webster's Unabridged Dictionary defines the verb *plagiarize* as "to take and pass off as one's own (the ideas, writings, etc., of another)." To put it simply, if you use an author's exact words, be sure to put them in quotes, for example if Ms. XY says in her book entitled *Flea Flicking Made Simple*, "the most efficient way to flick fleas is using the thumb and forefinger." If you don't use an exact quote but very closely paraphrase the author, you could say: Ms. XY tells us, in *Flea Flicking Made Simple* that using the thumb and forefinger is a most effective method.

Another way of crediting the author is by using a footnote. You could state in your paper: Flea flicking is best accomplished with the thumb and forefinger[1], where a footnote numbered 1, citing Ms. XY's book would appear on the bottom of the page.

As you progress in the academic world, plagiarism is considered an extremely serious offense, since it is, in a sense, taking someone else's original ideas or research and claiming it for your own. Therefore, be sure to get into good habits of crediting your sources, right at the start.

Just as with the bibliography, footnotes follow a specified format.

WRITE AND REWRITE

Your rough draft is just that, a first attempt at writing your research paper. There are no hard and fast rules as to how many drafts a paper should go through, but this is the opportunity to correct, revise, rearrange and reword your material to its best advantage. You may wish to type or print your early drafts with triple spacing, to give yourself enough room to write in many corrections.

Have family members and friends review your work too. Although they may not be scientists or writers, they can find spelling errors, awkward grammar, or badly worded sentences that you are too close to spot. Keep your handy dictionary and thesaurus at your side as you write.

A. Periodical
   Patrick Huygne, "Earthquakes, the Solar Connection" *Science Digest,* XC, (October, 1982), 73-75
B. Interview
   Dr. J. Holman III, Orthodontist, *Interview,* 11/23/85
C. Books
   1) by single author:
      Lauber, Patricia, *Of Man and Mouse.* New York: The Viking Press, 1971
   2) by multiple authors:
      Anderson, Garron P., S. John Bennett and K. Lawrence DeVries, *Analysis and Testing of Adhesive Bonds.* Long Beach, CA: Foster Publishing Co., 1971

D. Encyclopedias
   1) Selection without author
      "Galena," *Encyclopedia International,* 1974
   2) Selection with author
      Roderick, Thomas H., "Gene," *Encyclopedia International,* 1974

E. Government or other institutional publication
   Science Service. *ABSTRACTS - 37th Internation Science and Engineering Fair,* Washington, D.C., 1986

Fig. 4-8. Final bibliography.

a good impression on science fair judges.

Regardless of how it was created, be sure your margins are correct, your text is double-spaced, your name is on the paper and you have obeyed any formatting rules your teachers have given you. Remember to include your bibliography, in finalized format, at the end of the report.

Computer Note:

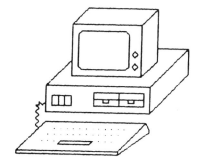

This may be almost too obvious to mention. Those of you with computers can throw away the erasers, correction tape, and white fluid, and just use your keyboard.

If you do not have a spelling checker, this may be the time to get one. These programs are quite inexpensive, and will give you valuable assistance throughout your academic career. Some of them even include a thesaurus, which will help you to find synonyms for overworked words.

When you are satisfied that your paper is the best that it can be, you are ready to create your final copy. If, until now, your drafts have been handwritten, try to borrow a typewriter. Even if your handwriting is good, your teacher will appreciate seeing typed copy, and it will certainly make

Computer Note:

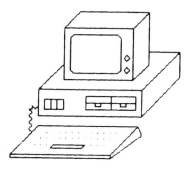

If you have been using a computer, this, of course, is simply printing out your latest version of the report. Now that this phase of the project is complete, remember to make a backup diskette. Be sure that all your files, the notes, bibliography, and footnotes, as well as the final research paper, are included. This is your insurance policy!

Chapter 5

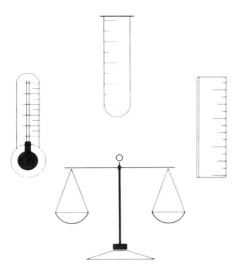

The Question and the Hypothesis

"Why is there air?" asks Bill Cosby, at the beginning of one of his famous routines. Although you would hardly ask such a broad or general question as the basis of a science project, (especially since you could never hope to finish the experiment in time!) the objective of *all* science is to answer a question or solve a mystery concerning the world around us.

The experiment you'll do, however, should try to solve a very specific problem. When you are deciding on the question (or purpose of the experiment) be sure that it's one that you can answer, with the time, resources, skills, and equipment available to you. It therefore must zero in on a limited aspect of your general topic. When focusing on your main purpose, keep in mind exactly how you plan to carry out your experiment and what you need to do it. After you consider these factors you may change the question you've selected.

The information you gathered while you were writing your research paper should come in handy now, helping you to focus on the purpose of the project, and pose that specific question. Remember, however, that an effective project cannot be based on an inquiry that can be answered with research alone. Your problem should be one you can solve by experimentation, followed by analysis of the results.

Notice that we've used the word *specific* several times. The biggest mistake beginners make at this point in the project is to define their topic too broadly.

You will notice that the questions in Fig. 5-1 are very general. They are vague, and do not indicate the substances or quantities to be tested, or how differences will be measured.

Instead, formulate your question, identifying exactly what your experiment will test and the comparisons you

intend to make. Examine Fig. 5-2 and notice that the same topics used above have been narrowed down and reworded into effective questions that express the problem more specifically.

This set of questions indicates what will be compared and what measurements will be used to show similarities or differences. For example, question 3 not only states what we are attempting to accomplish with the experiment, but also what is to be measured.

When you carefully read and compare all five questions on both lists, you'll see that the best questions for a science project are those that are extremely limited in scope. They state the intent of the experiment, list the subjects to be tested, and then describe results to be measured.

Once you have decided on your question, you must develop your *hypothesis* before designing and executing your experiment. You probably have a good guess, or at least an opinion, on what you think the answer to your question will be. You may have held this theory before doing your research, but after gathering the data, this should be an educated guess, based on the information you've learned so far.

Let's stop here for a moment now, and be sure that we understand all the terms. A *fact* is something that has been proven true. An *opinion* is something you believe to be true. Webster's Ninth New Collegiate Dictionary defines an opinion as: "a belief stronger than an impression and less strong than positive knowledge." The hypothesis, on the other hand, is more than an opinion and less than a fact. It represents your prediction of what the results will be, or, in other words, your *opinion* based on the *facts* that you've learned.

The experiment will be a test, or series of trials, which

1. Does television have a bad affect on test scores?
2. Does sound affect appetite?
3. What plants are flame retardant?
4. What is the effect of irrigating plants with salt water?
5. Do teenagers get overexcited watching MTV?

Fig. 5-1. Poorly worded questions.

1. What is the difference in test scores between sixth graders watching less than 3 hours of TV per night and 3 or more hours of TV per night?
2. What is the difference between the appetite in mice subjected to high and low frequency sounds, or no sound at all?
3. What are the rates of combustion of five different types of plants growing in the canyon areas of San Diego?
4. What is the difference in growth rate of three different types of plants irrigated with tap water and the same plants irrigated with sea water?
5. What is the difference in pulse rate between eighth graders watching MTV and listening to classical music?

Fig. 5-2. Well worded questions.

1. I believe that the test scores of 6th graders who watch more than 3 hours of TV per evening will be at least 20% lower than those watching less than 3 hours, because children watching more TV spend less time reading.
2. I believe that the amount of food eaten by mice subjected to high and low frequency sounds will differ from the amount eaten by those not subjected to those sounds.
3. I believe that iceplant and other succulents burn more slowly than other canyon plants, such as mesquite, sage, and giant dandelions, due to the greater degree of stored moisture.
4. I believe that sea water will retard growth in house plants such as spider plants, pothos, and Boston Fern, due to the high salt content, which is toxic to plants.
5. I believe the pulse rate in eighth graders watching MTV will be 50% higher than the pulse rate of those listening to classical music, because MTV is louder and faster.

Fig. 5-3. Hypotheses.

Mendelevia, he was not attempting to prove a hypothesis, but rather to create a computer game that would make a boring subject fun to learn.

Although you will not need a complete list of variables and controls at this time, you will need at least a rough idea of these in order to formulate your hypothesis. Briefly, a variable is something to be changed during the experiment and a control is something to be held constant, against which measurement is made. As shown in the examples, your

you will perform in order to prove your theory. You will apply various substances or use different quantities of something on a particular subject, then measure the variations, and finally record results. Therefore, your statement must identify *what* you are testing, *how* you are testing it and *why* you expect the results to prove your hypothesis. Figure 5-3 gives examples of well-worded hypotheses that relate to the questions asked in Fig. 5-2.

You'll notice that all the hypotheses begin with the words "I believe . . ." This indicates that it is only a theory until the experiment has been done and its results analyzed.

Sometimes a hypothesis will contain not only your theory of what the experiment will prove, but why you're making this assumption. In Andrew Wolf's experiment testing the porosity of various types of Hohokam pottery, he assumed that cooking required pottery of different porosity than storage.

In certain types of projects, the hypothesis is not so much a theory as a statement of what the experiment hopes to accomplish. In William Chen's project BLACKJACK, his work did not attempt to prove a theory, but instead, tried to improve different blackjack systems available. When Angel de la Cruz developed his computer program

hypothesis should mention:

—The subject of your experiment
—The variable to be changed
—The variable to be measured
—The results you expect

The subjects of the first experiment are 6th grade students. The measured variable is the test scores, and the experimental variable is the amount of television watched. The result expected is that the students watching more TV will have lower scores. In the third example listed, the subject to be tested is brushfire, the variable to be tested is the type of plant, the variable to be measured is the rate of burning, and the results you expect are that succulent plants burn more slowly than nonsucculent plants. In the fourth example, the variable to be changed is the type of water used, the controls are the type of plant and amount of water, and the measured variable is the amount of growth.

Although the Question and Hypothesis are the shortest pieces of writing you'll do during your project, they're vitally important. They form the basis of the entire experiment, and are the basis on which your results are judged.

Chapter 6

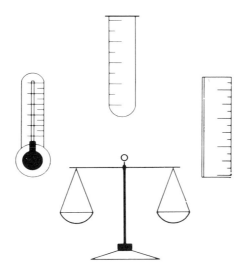

The Experiment

The heart of your science project is the experiment. To be effective and successful, the experiment must be carefully and methodically planned. Science fair judges on all levels overwhelmingly agree that the most important factor in rating your entire project is the proper use of the scientific method.

The scientific method is an organized system of conducting an experiment, including the collection, measurement and documentation of data. It must include a specified variable and control, as well as at least one experimental and one control group. In addition, the experiment must use a large enough number of samples, or perform sufficient repetitions of the test, to assure a valid result.

Let's begin by defining the terms *subject, variable*, and *control,* and how they are used.

SUBJECT OF THE EXPERIMENT

Any valid experiment is designed to test and examine the effect of a change in environment or condition. To conduct such a test, there must be a specific subject, one or more variables, and a control. The subject of the experiment is the item being tested. For example, in John Stoffel's project, *The Effect of Radiation on Bean Seeds and Seedlings,* the bean seeds and seedlings are the subjects of the experiment.

The rabbits are the subject of Elizabeth M. Robinson's project entitled *The Rabbit Test,* in which she tested the effects of steroids on the reproductive rates of rabbits.

VARIABLES

One variable in your experiment is the factor that is altered. This is often the substance or condition that your project is studying. For example in Nicola Kean's project *What happens when the length of a Pendulum is increased?* the length of the pendulum is the variable. In an experiment that concerned radiation and plants, conducted by Jeffrey S. Morgenheim, the amount of laser radiation was the variable. These are examples of what is called either the *experimental variable,* or the *independent variable.*

Another type of variable is the *measured variable,* which is also known as the *dependent variable.* This represents what you are evaluating, such as the amount of growth in a particular specimen. Daniel S. Chang, of Pago Pago, American Samoa conducted an experiment to determine whether carbon dioxide was necessary for coral growth. The amount of growth in his coral specimens was the measured variable.

Let's examine some of the projects we discussed in Chapter 5 and look at the subject as well as the independent and dependent (or experimental and measured) variables shown in Table 6-1.

CONTROLS

Once the variables have been determined, it's important to list the things that must not be allowed to change. These are known as the *controls,* or *controlled variables.* (Don't confuse this term with "control group," which we'll discuss in a few minutes.) Controls are applied to *all* subjects being tested. If your experiment changes in any way, other than the stated variables, it will be impossible to determine whether your intended variable or other dissimilar factors caused the results that occurred. For example, if testing the effects of caffeine on different groups of goldfish, the amount and type of food given to the groups, as well as the lighting

Fig. 6-1. John Stoffel; *The Effect of Radiation on Bean Seeds and Seedlings.*

available to the fish in the separate tanks, must be identical. Otherwise, it would be impossible to prove whether differences in the groups at the end of the experiment were caused by varying amounts of caffeine, or the alterations in diet, lighting, etc. Notice the controls for our famous five projects shown in Table 6-2 will keep all the factors except the independent variable identical throughout the experiment.

EXPERIMENTAL AND CONTROL GROUPS

In order to make a statement about the results of your experiment, you must compare the results of applying your variable to your subject with your subject under normal conditions. In other words, what happens in the course of your experiment must be compared to the condition of your subject before anything was changed. To accomplish this you will need to divide your subjects into *at least two groups.*

Your *experimental group* is a collection of subjects to which the *experimental variable* is applied. If you have multiple independent variables, apply one at a time. The reason for this is that if you change several variables simultaneously, you'll never be sure exactly what caused your results. Table 6-3 lists experimental and control groups.

It is also very important to be sure that all the subjects used in both groups are identical. For example, in Daniel S. Chang's project, which determined whether carbon dioxide was necessary for coral growth, the most difficult part of the project was collecting identical coral specimens for the experimental and control groups.

Another way of applying multiple variables, or several forms of the same variable, in order to carry out the purpose

Table 6-1. Subjects and Variables.

Subject	Experimental Variable	Measured Variable
Sixth graders	Hours of TV watched	Test scores
Mice	Type of sound	Appetite
Brush fire	Species of plant	Rate of combustion
House plants	Type of water	Amount of growth
Eighth graders	Variety of music heard	Heart rate

Fig. 6-2. Elizabeth M. Robinson; *The Rabbit Test; Phases V and VI.*

Table 6-2. Project Controls.

Subject	Project Controls
Sixth graders	The type of courses taken, the other environmental factors, such as whether they have their own room.
Mice	Amount of light and water each group gets.
Brush fire	Canyon areas where plants are native.
House plants	Amount of light and plant food each specimen receives.
Eighth graders	Amount of aerobic and other activities the Students indulge in.

of your experiment, is to use a number of experimental groups. In Randall Shank's experiment, *NutraSweet ™ In Diet Soda: A Study of Storage Temperature* there were four experimental groups, stored at different temperatures, 5°, 15°, 25°, and 35° C. In another project studying popular soft drinks, Jennifer L. Decker tested the sugar content in eight different regular and diet soft drinks, using glucose solutions and testing for the presence of sugar.

To make your experiment manageable, restrict the number of variables or variations. In other words, if you limit the factors you're changing and hence, the number of experimental groups you will use, you'll keep your squash plant experiment from being the killer gourd that ate New Jersey.

The *control group* is a collection of subjects, identical to those in the experimental groups studied, to which no variables are applied. For example, if you are testing the effect of crushed bone meal added to the soil for tomato plants, your control group would consist of a group of tomato plants that *did not have* bone meal added to the soil. With the exception of not applying the stated variable to the control group, it must be *exactly* like the other group(s) in all other respects, such as type, number, environment, brand, etc. Otherwise, the comparison, and hence, the results of your experiment, may be invalid.

For example, in Libby Haines' project, *Can A Charcoal Filter Reduce the Effects of Automobile Exhausts on Chick Embryos?* the control group consisted of embryos which were not subjected to any fumes, as opposed to the experimental groups, which were subjected to filtered and unfiltered fumes.

PROCEDURES

The first step in designing your project is to detail the exact experimental procedures. It should be a step-by-step list that anyone, even a person who knows nothing about science, could follow in order to duplicate the experiment. Each item in the list should describe only one action, in the exact sequence that the particular step is performed in the course of the experiment. Because you are breaking the

Table 6-3. Experimental and Control Groups.

Subject	Experimental Group(s)	Control Group
Sixth graders	Those that watch more than 3 hours daily	Those that watch less than 3 hours daily
Mice	Those that are subject to periods of sound	Those not subjected to periods of sound
Brush fire	Burning of succulent canyon plants	Burning of non-succulent canyon plants
House plants	Group watered with sea water	Group watered with tap water
Eighth graders	Group listening to MTV	Group listening to classical music

---

Procedures

1. Buy 30 mice.
2. Buy cages and food.
3. Put 10 mice in each cage and put them in separate rooms.
4. Give mice the same amount of food each day.
5. Test two of the groups with sound.
6. Measure and record the amount of food that each group eats.

Fig. 6-3. Poor procedures.

procedure down to such thorough detail, it may contain several pages, including not only the experiment itself, but the steps taken to buy, borrow, or build your materials, set up the experiment, and finally, conduct the necessary tests.

Probably, your advisor or science teacher will want to review your procedures before you actually begin. He or she will examine them very critically, looking for specific, detailed, sequential steps. Next, the teacher will review each step. If your sample is too small, or you plan to conduct too few tests, he or she may ask you to add steps or describe them in more detail.

"But I meant that," you might protest. "Anyway, that's really included in step 4." It may be obvious to you, but would probably be a complete mystery to that theoretical nonscientist who should be able to duplicate your experiment.

Study Fig. 6-3. It shows a list of procedures for a well thought-out experiment involving the correlation between sound and mouse appetite. However, the steps are too vague for anyone who's uninformed on the topic.

However, by simply spelling out every step in detail, as shown in Fig. 6-4, the list becomes an effective statement of procedures.

If you are using live vertebrate specimens, such as mice, show the steps you'll be taking to insure the adequate care and humane treatment of your subjects. Get the necessary permissions of parents, teachers, or other authorities.

Also, if your project uses tissues, organs, human, or animal parts, you will need to complete a form certifying the source of your tissue sample.

MATERIALS

Not only must the steps be precise, but they must include a materials inventory, which enumerates *everything* you will use. Again, this should be a list that is so specific that anyone could follow it. List the brands, sizes, quantities, contents and temperatures of all items that are necessary to conducting the experiment. Figure 6-7 compares vague and specific materials lists for the same project.

If there is anything you are using that you cannot describe verbally, illustrate it by including diagrams or photographs, making certain to include specific details of each item.

Some science students build their own machines or containers to use in their experiments. If you've done so, include instructions on how to construct it. These directions should be similar to (or, hopefully, better than) those that are included when buying a bicycle or a barbecue that requires assembly. They should include a parts list, as well as a description (and perhaps an illustration or photograph) of the finished product. Here too, be sure to list *specifically* all materials used. The instructions should consist of a numbered, sequential list, detailing what parts are used, in which order, to construct the device. Rather than something "so simple that even a child of four could do it," strive for a set of instructions that are easy enough for an ordinary parent to follow!

Procedures

*1. Buy 30 mice, 3 cage kits, rodent bedding material and mouse food from ABC Pet Shop.
*2. Buy a high frequency buzzer and a low frequency buzzer.
3. Set up each cage in a different room of the house.
4. Put 88 cubic centimeters of bedding, 1.70 cubic centimeters of food and 60 cubic centimeters of water in each cage.
5. Put 5 male and 5 female mice in each cage, and label the cages A, B and C.
6. Wait 3 days to begin experiment to allow mice to get adjusted to surroundings. During that period, measure amount of food and water consumed.
7. Each day during experimental period (11/18 through 2/2):
   a. In each cage, measure the amount of food left in the food dish. Subtract the amount from the amount given each morning (1.87 cubic centimeters) to determine the amount eaten, which is recorded in the experimental log.
   b. Fill food dish in each cage with 1.87 cubic centimeters each morning. (The amount consumed during pre-test periods plus 10%)
   c. Refill water bottle in each cage.
   d. Each day, expose the mice in group B to the low frequency sound for 1 hour and expose the mice in group C to the high frequency sound for 1 hour.
8. Each Saturday during the experimental period, move the mice in each cage to a large cardboard box (together with their food dish) while cages are cleaned and the cedar bedding is replaced.
9. When the experiment is completed, re-sell the cages and mice back to ABC Pet Shop.
*For details, see materials list.

Fig. 6-4. Good procedures.

CERTIFICATION OF HUMANE TREATMENT OF LIVE VERTEBRATE ANIMALS

Name of Entrant _____ School _____

Project Title _____

Any student research involving animals **MUST COMPLY** with the requirements of the **California Education Code** stated here:

HUMANE TREATMENT OF ANIMALS, State of California Education Code Title 2, Division 2, Part 28, Chapter 4, Article 5

51540. In the public elementary and high schools or in public elementary and high school sponsored activities and classes held elsewhere than on school premises, live vertebrate animals shall not, as part of a scientific experiment or any purpose whatever:

    (a) Be experimentally medicated or drugged in a manner to cause painful reactions or induce painful or lethal pathological conditions.

    (b) Be injured through any other treatments, including, but not limited to, anesthetization or electric shock.

Live animals on the premises of a public elementary or high school shall be housed and cared for in a humane and safe manner.

The provisions of this section are not intended to prohibit or constrain vocational instruction in the normal practice of animal husbandry.

Compliance with the following ISEF regulations is **ALSO REQUIRED**; however, **THE PROVISIONS OF THE CALIFORNIA EDUCATION CODE MUST BE FOLLOWED WHENEVER CONFLICTING REGULATIONS OCCUR.**

EXCERPTS, INTERNATIONAL SCIENCE AND ENGINEERING FAIR (ISEF) REGULATIONS FOR EXPERIMENTS WITH ANIMALS

Prepared by Science Service, Inc., for exhibitors at the ISEF and Affiliated Fairs

1. The basic aims of experiments involving animals are to achieve an understanding of life processes and to further knowledge. They do not include the development of new or refinement of existing surgical techniques or experiments in toxicological studies. Experiments involving animals (live or preserved, vertebrate or invertebrate excluding **Homo sapiens**), vertebrated embryos and fetuses, and embryos of fowl within three days of hatching, must have clearly defined objectives requiring the use of animals to demonstrate a biological principal or answer scientific propositions. Such experiments **must** be conducted with a respect for life and an appreciation of humane considerations.

2. The use of protista and other invertebrates is to be encouraged for most research involving animals. Their wide variety and the feasibility of using larger numbers than is usually possible with vertebrates makes them especially suitable.

3. To provide for humane treatment of animals, an animal care supervisor knowledgeable in the proper care and handling of experimental animals **must** assume primary responsibility for the conditions under which the animals are maintained. If the school faculty includes no one with adequate training in this area, the services of a qualified consultant **must** be obtained.

4. All live or preserved animals or animal parts **must** be lawfully acquired from an approved source and their care and use **must** be in compliance with local, state and federal laws.

5. The **comfort** of the animals shall be a prime concern. No research using live vertebrate animals shall be attempted unless the animals are obtained from a reliable source and the following conditions can be assured: appropriate, comfortable quarters; adequate food and water; humane treatment and gentle handling. Care must be provided at all times, including weekends and vacation periods.

Fig. 6-5. Certification of humane treatment of live vertebrate animals.

GSDSEF ANIMAL REGULATIONS

1. Experiments involving procedures not in violation of the "painful reaction" or "injured" restrictions of the California Education Code are permitted if certified by a qualified biomedical scientist PRIOR TO the beginning of the investigation. (NOTE: GSDSEF rules do not permit students or their adult supervisors, as part of a student-planned project, to 1) perform surgery; 2) conduct experiments involving toxicity, nutritional deficiency or harmful physical or psychological stress or 3) perform the sacrifice of live vertebrate animals.

2. It is permissible for the student and designated adult supervisor to consult with a biomedical scientist to obtain detailed instructions and guidance in techniques to be used by the student under the direct continuous supervision of a designated adult supervisor (for research not conducted in the biomedical scientist's lab). In this instance the designated adult supervisor will be required to certify in writing jointly with the biomedical scientist.

3. Either the biomedical scientist or adult supervisor must provide continuing supervision to assure compliance with the protocol.

4. Major deviations from the approved protocol may be implemented only with the written approval of the biomedical scientist.

5. The biomedical scientist or adult supervisor must be in the same locality as the student for the duration of the experimental work except for short trips. This means that a project started in one city may not be continued in another unless an alternate designated adult supervisor, approved by the biomedical scientist prior to the continuation of the experimental work, agrees to supervise the project.

6. A biomedical scientist is defined as one who possesses an earned doctoral degree in science or medicine and who has current working knowledge of the techniques to be used in the research under consideration.

7. A designated adult supervisor is defined as an individual who has been properly trained in the techniques and procedures to be used in the investigation. The biomedical scientist must certify that the designated adult supervisor has been so trained.

RESEARCH PLAN

Purpose of Project:

Starting Date: _____

Site at which investigation will take place:

    Name _____

    Address _____

Live vertebrate animals to be used:

    a) Genus, species and common name _____

    b) Number of animals _____

    c) Animals obtained from _____

List objectives of the experiment and describe fully the methods and techniques involved. When the use of electrical current, laser beams, sound stimuli or other artificial stimuli are an integral part of the Research Plan, they must not exceed the normal tissue tolerances for the species concerned (AS INDICATED IN THE **Biology Data Handbook,** 2nd Edition; editors, P.O. Altman and D.S. Dittmer; publisher, Federation of American Societies for Experimental Biology).

Describe proposed methods of animal care:

    a) Cage size _____

    b) Number of animals per cage _____

c) Temperature range (maximum and minimum) in degrees Celcius of room where animals are to be kept.

_____

d) Frequency of feeding and watering _____

e) Frequency of cleaning cage _____

f) Type of bedding to be used _____

g) Where will animals be housed? _____

h) Where will animals be returned when research is complete?

_____

Name of Animal Care Supervisor _____ Name of Biomedical Scientist _____

Name of Designated Adult Supervisor _____

Signature of Student _____

CERTIFICATIONS
THE FIRST TWO CERTIFICATIONS MUST BE COMPLETED FOR ALL PROJECTS INVOLVING LIVE VERTEBRATE ANIMALS

CERTIFICATION BY TEACHER/ADVISOR   I agree to sponsor the student named above and assume responsibility for compliance with the existing rules and regulations pertaining to experiments with animals.

Signature _____

Name (type or print) _____   Date _____

Institution _____   Title _____

Institution Address _____   Phone _____

Home Address _____   Home Phone _____

CERTIFICATION BY ANIMAL CARE SUPERVISOR   of compliance with California Education Code and Paragraph 3 of the ISEF Regulations for Experiments with Animals. (Must be completed prior to receipt of animals by
the student.)

> I certify that I have reviewed and approved the Research Plan and will supervise and accept primary responsibility for the quality of care
> and handling of the live vertebrate animals used by the designated student. I further certify that I am knowledgeable in the proper care
> and handling of experimental animals and meet prevailing animal supervisory requirements.

Signature _____

Name (type or print) _____   Date _____

Institution _____   Title _____

Institution Address _____   Phone _____

Home Address _____   Home Phone _____

This section MUST be completed for ALL vertebrate animal projects.

Fig. 6-5. Continued.

Complete this section for all projects except observations of animals in their natural environment.

CERTIFICATION BY BIOMEDICAL SCIENTIST (if required) of compliance with the California Education Code, the ISEF Regulations for Experiments with Animals and the GSDSEF Animal Regulations. (Must be completed prior to the start of the project.)

I certify that I have reviewed and approved the Research Plan; that if the student or designated adult supervisor is not trained in the necessary procedures I will ensure his/her training; that I will assure that the requirements of the California Education Code, the ISEF Regulations for Experiments with Animals and the GSDSEF Animal Regulations are fully met; that I will provide advice and supervision personally or through a designated adult supervisor throughout the project; and that I am a qualified scientist with an earned doctoral degree (Ph.D., M.D., D.V.M.) and a working knowledge of the techniques to be used by the student in this research.

Signature _____

Name (type or print) _____ Date _____

Institution _____ Title _____

Institution Address _____ Phone _____

Home Address _____ Home Phone _____

CERTIFICATION BY DESIGNATED ADULT SUPERVISOR (if required)

I certify that I have been trained in the techniques to be used by this student and will provide direct supervision for the research.

Signature _____

Name (type or print) _____ Date _____

Institution _____ Title _____

Institution Address _____ Phone _____

Home Address _____ Home Phone _____

This form, properly completed, must be part of the carefully planned procedures for experimentation with live vertebrate animals or animal parts. It must accompany any such project exhibited at, or when presented for any public display associated with the Greater San Diego Science and Engineering Fair. All sections must be completed except the "Certification by Biomedical Scientist" and the "Certification by Designated Adult Supervisor" which need be completed only by those students carrying on experiments other than observations of animals in their natural environment.

Form GSDSEF-2 (1987)

CERTIFICATION OF TISSUE SAMPLE SOURCE

Name of Entrant _____ School _____

Project Title _____

This form must be completed for all projects using tissue(s), organ(s), human part(s), or animal parts, including blood.

When live or preserved tissue samples or parts of human or vertebrate animals are obtained by the student from an institution or biomedical scientist, a statement signed by the adult providing the tissue is required. Students may NOT be involved in the direct acquisition of these samples from living human or vertebrate animals.

Live tissue samples must be:

    a) from a continuously maintained tissue culture line already available to institutional researchers.
       or
    b) from animals already being used in an on-going institutional research project.

RESEARCH PLAN

1. Tissue(s), organ(s), or part(s) used: _____
                                                         _____
                                                         _____

       Tissue sample is from: (check appropriate box)
       ☐ Human Source
       ☐ Vertebrate animal source
       Genus, species and common name _____

2. Starting Date: _____

3. Purpose of Project:

Fig. 6-6. Certification of tissue sample source.

4. List objectives of the experiment and describe fully the methods and techniques involved:

Signature of Student _____ Date _____

CERTIFICATION

Institution or company that is source of Tissue Sample:

Name _____

Address: _____

_____

I certify that the above listed materials were provided by me or my institution and that the student listed was NOT involved in the direct acquisition of the samples provided or purchased.

_____   _____
             Signature   Title

_____   _____
               Date   Telephone

Form GSDSEF-4 (1987)

Vague	Specific
1. 30 mice	1. 30 house mice, 15 of each sex.
2. 2 buzzers	2. One piezzo sound buzzer of high frequency (3800 cycles per second).
3. 3 cage kits	3. One piezzo sound buzzer of low frequency (1500 cycles per second).
4. Mouse food	4. Three identical cages measuring 11 × 16 × 2 inches each.
5. Cedar bedding	5. Three identical play wheels.
6. Food dishes	6. 7 kg. mouse food, consisting of corn, sunflower seeds, wheat and oat grain.
7. Water bottles	7. 975 cubic centimeters of cedar chip bedding.
	8. Three identical water bottles.
	9. Three identical food dishes, each with a volume of 3.4 cc.

Fig. 6-7. Materials list.

Many extraordinary projects can be done using very ordinary materials. However, if you require special equipment or supplies, you may need some help in locating these, or else you may need to alter your experimental design. Before deciding to build or purchase what you need, find out if you can borrow the equipment from your school laboratory. Teachers may permit you to use it after school hours, or even allow you to take some items home. However, first investigate what the fees are for any loss or breakage. You may have second thoughts about using these things without supervision.

Other possibilities are university or corporate lab facilities. This has been a particularly good resource for those with a mentor or advisor who works in a laboratory. Several students with successful projects have developed excellent working relationships with scientists who have helped them. Sometimes, friendships have developed that extended far beyond the life of the science project.

It may turn out that your only course of action is to buy your materials. If so, see if you can buy these through your school in order to take advantage of discounts that may be available to a large school district. If not, you will have to contact various supply houses for the live specimens, chemicals, or equipment you may need. The best place to begin is your local *Yellow Pages,* for Chemicals, Biological Supplies, Laboratory Equipment, Plant Nurseries, etc.

If your experiment is concerned with rare, or exotic

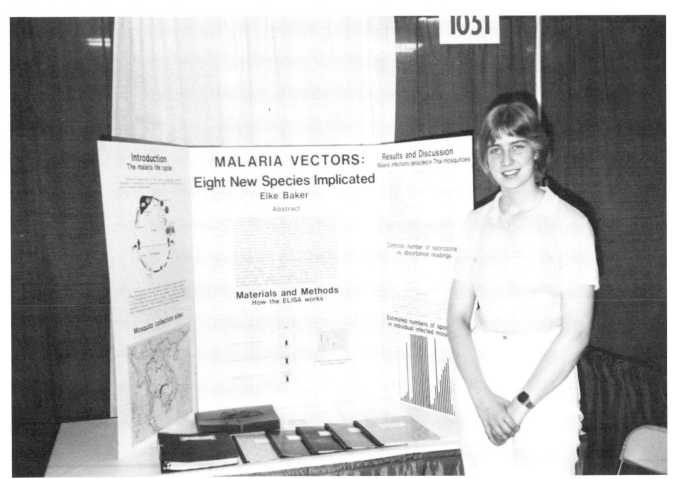

Fig. 6-8. Elke Baker; *Malaria Vectors—Eight New Species Implicated.*

Table 6-4. Sample Size.

Subjects	Suggested number of subjects per group
Plants	10-50
Live Vertebrates	10-15
Humans	50-100

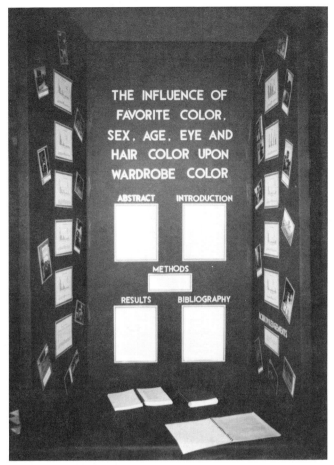

Fig. 6-9. Lan Do; *The Influence of Favorite Color, Sex, Age, Eye and Hair Color upon Wardrobe Color.*

someone drinks the soda, or the baby used "it" to make mud pies, that's what!

Obviously, when dealing with live specimens, a larger sample size is more important, because it may be almost impossible to determine whether the individuals in the group were average or abnormal. If you're dealing with seedlings, start with at least 50, and preferably more, to account for the fact that many will not germinate, or may later die.

Certainly, when dealing with human responses to a survey or questionnaire, you must include enough subjects to insure that your results are representative. For example, in Kimberly Lea Quinlan's project on *Teenage Stress* her results were based on 378 responses to a questionnaire. For the project *The Influence of Favorite Color, Sex, Age, Eye and Hair Color Upon Wardrobe Color,* Lan Do compiled the results from 419 surveys.

To make your results at all believable, you'll want to use at least 100 surveys. Incidentally, it's probably a good idea to distribute more than you need, to compensate for those who throw away your survey. Review the tables that follow as a guideline on sample sizes.

Regardless of the number of units, perform a sufficient number of tests or trials to make your results more certain. Statistically, no conclusive evidence can be inferred based on too few trials. Many teachers agree that a minimum of five trials is needed to prove a hypothesis, and even five is a bare minimum, which would be effective only if all the trials yielded the same result. To allow for the "worst possible case," ten trials would be a good minimum.

DOING THE EXPERIMENT

By this time, you probably have all your equipment, and can't wait to begin. Before you start, however, let's take a moment to reemphasize why you're doing this and what you hope to gain.

To achieve your scientific goals and objectives, you must know how to conduct an experiment. Consistency and regularity of procedures and measurements will insure that

specimens or substances, your search for the proper materials will be even wider. Elke Baker from Gaithersburg, Maryland, competed in the 1986 International Fair, with her project *Malaria Vectors: Eight New Species Implicated* in the Medicine and Health category. She found her specimens, eight various mosquitos, by working with the Walter Reed Army Institute of Research, and ultimately obtained samples from health agencies of the United Nations.

SAMPLE SIZE AND NUMBER OF TESTS

An important part of any experiment is the sample used. Common sense tells us that a procedure done *once* on only *one* subject will barely suggest a trend, much less prove a hypothesis. It is therefore important to have a large enough group to do adequate testing and provide conclusive information. Your results will be more credible if they happened to five subjects rather than to two, and will be just about conclusive if your sample size was ten. At the same time, you'll want to strike a balance and limit the number of elements to those you can adequately manage.

Another reason for having an adequate sample size is to include a number of extras, "just in case." In case what? In case the plant gets knocked over, the mouse escapes,

Table 6-5. Number of Trials.

Type of Project	Minimum	Suggested
Physics	20	50-100
Animal Behavior	10	25-50
Other	5	20-50

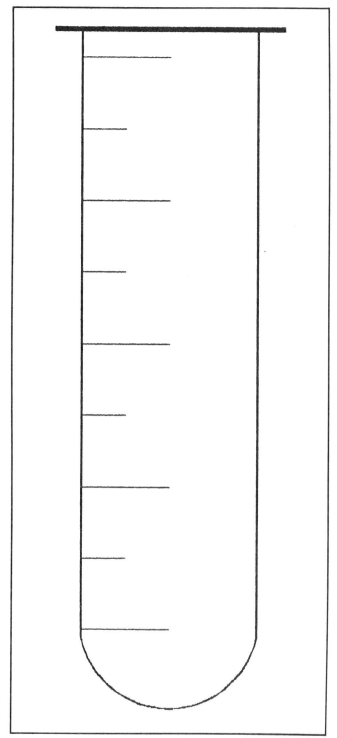

Fig. 6-10. Illustration of test tube.

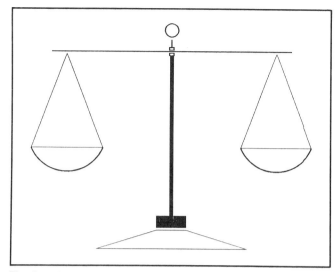

Fig. 6-11. Illustration of scales.

observation and computation is absolutely essential. First of all, show your professionalism by always using the metric system of measurement. Next, be certain that your measurements are exact. Finally, be sure that any instruments are precisely balanced or calibrated.

Wherever possible, you'll want to make a *quantitative analysis* of your results, based on true measurements, rather than simply a *qualitative analysis*, which relies on observation.

For example, "the mice kept in the dark ate twice the amount of food as the ones in the lighted room" is considered a qualitative analysis. In Gwen S. Pearce's project *Stop—You May be Using the Wrong Detergent,* the results

your data is correct. Understanding and using the experimental scientific method will increase your credibility with teachers and judges, giving them faith in what you've done, regardless of whether your experiment has "worked."

OBSERVATION AND MEASUREMENT

While you are doing your experiment, precise

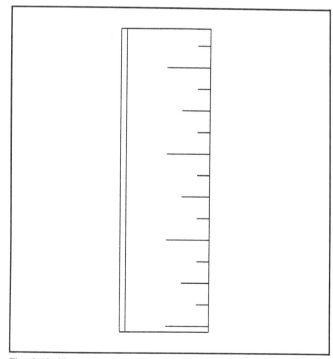

Fig. 6-12. Illustration of ruler.

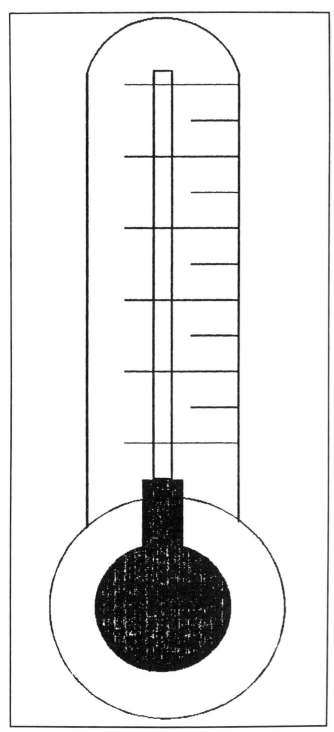

Fig. 6-13. Illustration of thermometer.

Testing Two Theories of Mouse Navigation she used the statistical T-test, thus insuring a quantitative analysis.

If you are using scales, rulers, vessels, or other items that measure weight, size, or volume, choose those that can show the smallest gradations you may require. "A little less than a milliliter," is not only vague, but cannot be compared to a quantity that is "almost a milliliter."

Often, the ability to measure the smallest gradations is essential to the project. For example John Chin-Hung Leu had to make minute, careful measurements of the amphibian eggs under various conditions in his experiment on the *Development of Amphibian Eggs under Simulated Hypogravity Conditions.* Therefore, use tools that will insure the greatest degree of accuracy and also will allow you to make exact comparisons.

RECORD-KEEPING

At the time that you make each and every observation, record your results in writing. This may well be the most important thing you'll do while conducting your tests or trials. A few minutes later, you might forget *exactly* what you saw or what calculations you made. Even if you have a photographic memory, your experiment will lose credibility if your recordkeeping is not current and accurate.

Logs and tables will help keep track of everything that happens during your project. Your tables will help you record the raw data you gather during your experiment. To make it as easy as possible, design a table before you even begin your experiment. Then, you can concentrate on carrying out your procedures and measuring your results without having to worry about "Where do I write that number?" or "One of the mice escaped. Where can I make a note of that?"

When designing your table form, include a place to record the date and time of each entry, as well as whatever measurements and observations you make. Also, leave a final column for notes or comments.

In Table 6-6, you could leave the ML (milliliter) designation out of each column, provided you noted somewhere on the page that all liquid volumes were expressed in milliliters. Note also that a comment such as "Found a hole in container B" would require a more detailed explanation in your diary or log. Such a condition might very well make your observations worthless, especially if you are measuring the amount of liquid that "disappears" in the course of your tests. When dealing with a control group and one or more experimental groups, it's a good idea to have a separate table for each group. This will help you to accurately record the information for each group.

While you're proceeding, don't forget to maintain your time log. This is extremely useful for whatever you do that can't be recorded on your table, for example, locating a graphing program, or conversations you've had with teachers or advisors. Include any additional research material you've read, or perhaps new discoveries you've made

of the cleaning tests with the various detergents were visually observed and evaluated, resulting in a qualitative analysis.

"Group A, kept in the dark room, ate between 8 and 10 cc per day, and group B, kept in the lighted room, ate between 4 and 6 cc per day," is an example of quantitative analysis, is based on exact measurement and is far more accurate. In Ann Wood's project, *Interaction vs. Backup:*

Table 6-6. Sample Experimental Log.

Date	Time	Liq. Absorbed Group A	Liq. Absorbed Group B	Comments
1/27	5:53 P.M.	5ML	5ML	Noticed a hole in container B

Table 6-7. Sample Experimental Log.

Date	Food Remaining (CC's)		
	Group A	Group B	Group C
12/5	6.9	4.8	4.8
12/6	7.1	5.3	3.6
12/7	7.3	5.0	3.9
12/8	7.0	4.8	4.3
12/9	6.8	4.7	4.1

while doing your project. Sometimes, in the course of an experiment, unforeseen problems may develop that cause you to change your procedures. Document any alterations, either in descriptive form, or using photographs, illustrations, sketches, charts, or graphs.

IF ALL ELSE FAILS . . .

For most of you, your experiment is going smoothly. You have measurable observations, which show at least some effects of your testing. If, however, things are going poorly, consider changing your procedures or altering your hypothesis. Jessica Wertlieb overhauled the experimental procedures seven times in the course of her project *Effects of UV Light on F'Plasmid of Escherichia coli,* when the microorganisms did not grow properly. If you started early enough and budgeted your time, you may even be able to change your topic if that becomes necessary.

If you're thinking of taking any of these steps, *please discuss it with your teacher or advisor.* First of all, she or he may realize that if you "hang in there" a little longer, the experiment might still yield some useful data. On the other hand, the teacher may see that by "not getting any results," you are, in fact, concretely disproving your hypothesis, which is equally valuable.

However, if your worst fears are realized and you cannot salvage your experiment, your teacher may be able to help you shift gears with a minimum of extra work. He or she will probably be able to guide you towards an alternate project, which can utilize some of your background research as well as the materials you've accumulated. In any event, you don't need to go it alone! If you've done your work and tried your best (your research, tables and logs will give ample evidence of that), there's always plenty of help available to you.

At this point, it is sometimes tempting to "fudge" your observations or measurements, especially if it appears that your experiment is completely disproving your hypothesis, or, worse yet, doing nothing at all. First of all, let me sympathize. Stuart Allen's experiment on dental adhesives "didn't work," and he was extremely discouraged to find that after all the research and preparation, absolutely nothing happened. He felt that, having "guessed wrong" the project was unsuccessful and he would receive a poor grade. However, because the project was well executed, it not only received a good grade, but he learned some important lessons about science.

Remember that "nothing happening" *is* a result. Rather than revealing the expected effect of the variable on the subject, your experiment may show that your subject is unaltered or minimally changed by the variable in the quantities you used. Although this was not the hypothesis you originally tried to prove, your experiment is equally valid, especially if you followed scientifically sound procedures and kept accurate records throughout. The tests may have refuted the hypothesis, which is as important as proving it.

Rather than disproving your theory entirely, the experiment may prove to be inconclusive. This usually shows that the variable was applied in the wrong quantity, strength, or temperature, or that more tests may be required. Look on the bright side. This gives you a built-in science project topic for next year!

Throughout scientific history, and the history of science projects, additional experimentation and research have turned failed projects into successful ones. Even experiments that work are often refined and enhanced to yield even more significant results.

Robin Rowe of Cocoa, FL, concerned about the prob-

Name			
Project Title			
			Time Log
Date	**Time**	**Activity**	
1/25/85	3:35 P.M.	One litter of 8 born in cage B. Removed mother and babies to a holding cage.	
2/05/85	4:30 P.M.	Experimented with Chart Master program to graph results.	
2/17/85	2:00 P.M.	Borrowed plotter and 6 different colored pens to draw graphs.	

Table 6-8. Time Log.

lems of water pollution in rural lakes, studied human and cow waste material that contaminated the water. Although she did not fully achieve her project goals, she entered the 36th International Science Fair (in itself, an accomplishment), and won a Third Award. The following year, Ms. Rowe applied the information she had learned about how to actually differentiate between the pollutants. Her project *Differentiating Between Human and Cow*

Escherichia Coli was entered in the 37th ISEF. She not only repeated her Third Place Award, but received a four year scholarship to Ohio State University.

As we mentioned earlier, more than the hypothesis or the results and conclusions, it is the experimental design and procedures, including meticulous observation and precise recordkeeping, which are the hallmark of a well-executed project.

Chapter 7

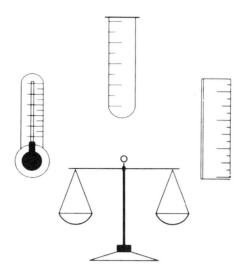

Results and Conclusions

Once the experiment is finished, you'll need to examine and organize your findings. Then, you'll interpret and analyze this data in order to formulate your conclusions.

RESULTS

Results, in terms of a science project, are defined as simply what happened during the course of an experiment. During this phase, the task is not to interpret the results. This will wait until you're ready to work on the conclusion, the true final phase of the experiment. What you'll want to do here is gather, organize, and reorganize the material in various ways, to make the results as clear and meaningful as possible.

The results phase of the project incorporates three elements, the raw data, the smooth data, and the analyzed data.

RAW DATA

The basis of the results will be the raw data, which consists of the tables you developed and maintained while conducting the experiment. It's really tempting to try to combine and analyze the information while you're actually doing the tests. However, it's vital that you simply record your observations. Anticipating the outcome may cause you to unconsciously bias your observations, which would, of course, invalidate the entire experiment.

You may wish to graph your raw results by creating bar and line graphs for your variable and control group(s), extending over the life of the experiment. For the project that we've been following, the effects of sound on mouse appetite, both the actual amount of food eaten and the percentage of available food eaten is shown in both tables and

bar graphs, as illustrated below in Fig. 7-1 and Tables 7-1 and 7-2.

In Mike Iritz's project *The Shaking Earth, The Burning Sky*, the incidence of high magnitude quakes, for the period between 1932 and 1974, was correlated to solar activity. Each factor was graphed separately, for a six year period as shown in Figs. 7-3 through 7-5.

SMOOTH DATA

Smooth data consists of the raw data that has been combined and correlated. For example, if your project examined one control and two experimental groups, whose measurements were charted and graphed daily, you will need to include the data over the life of the project into your smooth data. Figure 7-6 graphs the percentage of food eaten, as shown in Table 7-2 and Fig. 7-2, on a weekly basis.

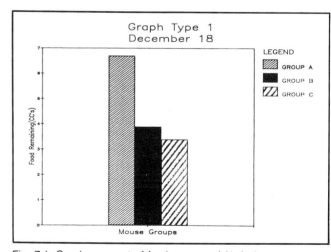

Fig. 7-1. Graph—amount of food consumed (1 day).

Table 7-1. Amount of Food Consumed.

Food Remaining (CC's)	Group A	Group B	Group C
Nov. 20	7.3	3.4	4.7
Nov. 21	6.2	4.3	3.9
Nov. 22	6.8	4.9	3.8
Nov. 23	6.6	5.0	3.7
Nov. 24	6.9	4.3	3.9
Nov. 25	7.1	4.6	4.0
Nov. 26	6.8	4.5	3.9
Nov. 27	6.6	4.1	3.8
Nov. 28	6.7	4.4	4.2
Nov. 29	6.4	4.7	4.1
Nov. 30	6.3	4.6	4.5
Dec. 1	6.5	4.9	4.3
Dec. 2	6.5	5.1	4.1
Dec. 3	6.7	4.7	4.2
Dec. 4	6.6	4.4	4.7
Dec. 5	6.9	4.8	4.8
Dec. 6	7.1	5.3	3.6
Dec. 7	7.3	5.0	3.9
Dec. 8	7.0	4.8	4.3
Dec. 9	6.8	4.7	4.1
Dec. 10	6.2	4.4	3.8
Dec. 11	6.5	4.3	3.5
Dec. 12	6.3	4.6	3.8
Dec. 13	6.7	4.8	4.3
Dec. 14	7.6	4.5	4.6
Dec. 15	7.2	4.1	4.0
Dec. 16	5.8	3.8	3.9
Dec. 17	6.4	3.4	3.8
Dec. 18	6.7	3.9	3.4
Dec. 19	6.9	4.6	3.6
Dec. 20	6.1	4.8	3.9
Dec. 21	6.1	4.7	3.3
Dec. 22	6.4	4.3	3.7
Dec. 23	6.6	4.2	4.2
Dec. 24	7.1	3.7	4.6
Dec. 25	7.4	3.9	4.9
Dec. 26	6.9	4.1	4.0
Dec. 27	6.4	4.3	3.6
Dec. 28	6.0	4.7	3.5
Dec. 29	5.9	5.1	3.2
Dec. 30	6.4	5.3	3.4
Dec. 31	6.7	4.7	3.8
Jan. 1	6.8	4.5	4.1
Jan. 2	6.4	4.8	4.3
Jan. 3	6.5	5.2	4.6
Jan. 4	6.3	4.5	4.0
Jan. 5	6.6	4.3	3.6
Jan. 6	6.9	4.0	3.7
Jan. 7	7.1	3.4	4.3
Jan. 8	7.3	3.7	4.7
Jan. 9	6.8	4.3	4.5
Jan. 10	6.6	4.6	4.2
Jan. 11	6.3	4.8	3.6
Jan. 12	6.5	4.2	3.5

In the earthquake and solar activity correlation project, the data from Figs. 7-3 and 7-4 were combined with the data shown in Fig. 7-5, and on the graphs shown in Figs. 7-7 and 7-8.

As shown in Figs. 7-9, 7-10 and 7-11, the earthquake data for both magnitudes studied, as well as the solar activity, was graphed for the 43 year period studied.

Finally, Figs. 7-12 and 7-13 show the correlation of each magnitude with the solar activity for the entire 43 year period.

If you find that you cannot include all your smooth data on one graph or table, you may use several, provided they combine the experimental and control groups.

Table 7-2. Percentage of Food Consumed.

Percentage of Food Ate	Group A	Group B	Group C
Nov. 20	85.4	93.2	90.6
Nov. 21	87.6	91.4	92.2
Nov. 22	86.4	90.2	92.4
Nov. 23	86.8	90.0	92.6
Nov. 24	86.2	91.4	92.2
Nov. 25	85.8	90.8	92.0
Nov. 26	86.4	91.0	92.2
Nov. 27	86.8	91.8	92.4
Nov. 28	86.6	91.2	91.6
Nov. 29	87.2	90.6	91.8
Nov. 30	87.4	90.8	91.0
Dec. 1	87.0	90.2	91.4
Dec. 2	87.0	89.8	91.8
Dec. 3	86.6	90.6	91.6
Dec. 4	86.8	91.2	90.6
Dec. 5	86.2	90.4	90.4
Dec. 6	85.6	89.4	92.8
Dec. 7	85.4	90.0	92.2
Dec. 8	86.0	90.4	91.4
Dec. 9	86.4	90.6	91.8
Dec. 10	87.6	91.2	92.4
Dec. 11	87.0	91.4	93.0
Dec. 12	87.4	90.8	92.4
Dec. 13	86.6	90.4	91.4
Dec. 14	84.6	91.0	90.8
Dec. 15	85.6	91.8	92.0
Dec. 16	88.4	92.4	92.2
Dec. 17	87.2	93.2	92.4
Dec. 18	86.6	92.2	93.2
Dec. 19	86.2	90.8	92.8
Dec. 20	87.8	90.2	92.2
Dec. 21	87.8	90.6	93.4
Dec. 22	87.2	91.4	92.6
Dec. 23	86.8	91.2	91.6
Dec. 24	85.8	92.6	90.8
Dec. 25	85.2	92.2	90.2
Dec. 26	84.2	91.8	92.0
Dec. 27	87.2	91.4	92.8
Dec. 28	88.0	90.6	93.0
Dec. 29	88.2	89.8	93.8
Dec. 30	87.2	89.4	93.2
Dec. 31	86.6	90.6	92.4
Jan. 1	86.4	91.0	91.8
Jan. 2	87.2	90.4	91.4
Jan. 3	87.0	89.6	90.8
Jan. 4	87.4	91.0	92.0
Jan. 5	86.8	91.4	92.8
Jan. 6	86.2	92.0	92.4
Jan. 7	85.8	93.2	91.4
Jan. 8	85.4	92.6	90.6
Jan. 9	86.4	91.4	90.8
Jan. 10	86.8	90.8	91.6
Jan. 11	87.4	90.4	92.8
Jan. 12	87.0	91.6	93.0

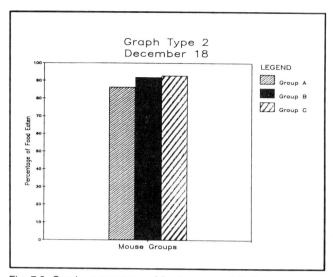

Fig. 7-2. Graph—percentage of food consumed (1 day).

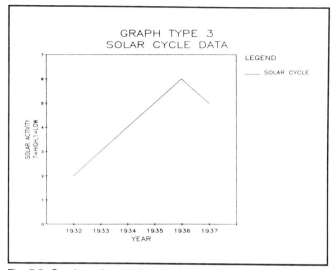

Fig. 7-5. Graph—solar activity—1932-1937.

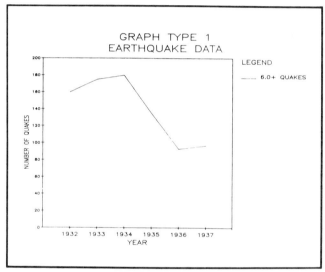

Fig. 7-3. Graph—6.0 earthquake activity—1932-1937.

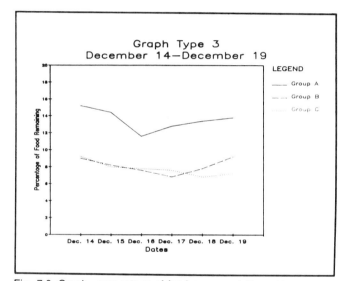

Fig. 7-6. Graph—percentage of food consumed (1 week).

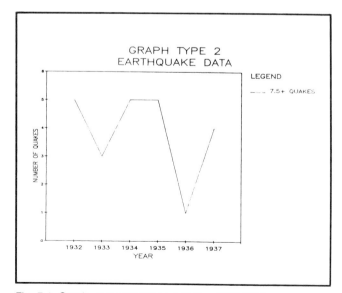

Fig. 7-4. Graph—7.5 earthquake activity—1932-1937.

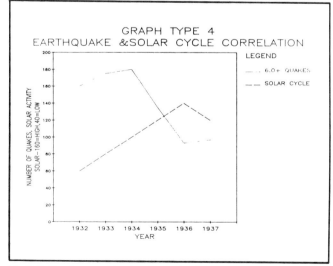

Fig. 7-7. Graph—combined solar and 6.0 earthquake activity—1932-1937.

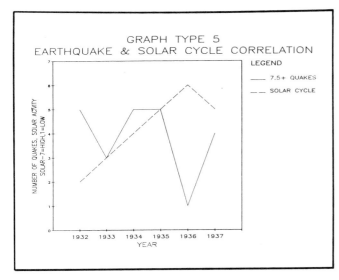

Fig. 7-8. Graph—combined solar and 7.5 earthquake activity—1932-1937.

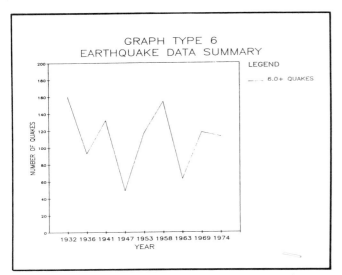

Fig. 7-9. Graph—6.0 earthquake activity—1932-1974.

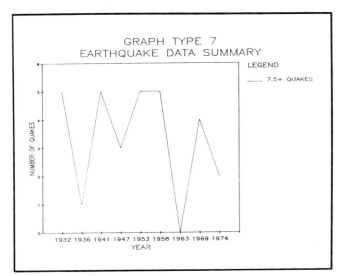

Fig. 7-10. Graph—7.5 earthquake activity—1932-1974.

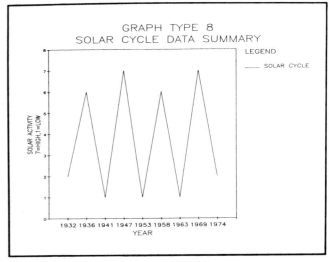

Fig. 7-11. Graph—solar activity—1932-1974.

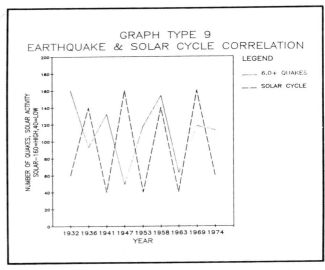

Fig. 7-12. Graph—combined solar and 6.0 earthquake activity—1932-1974.

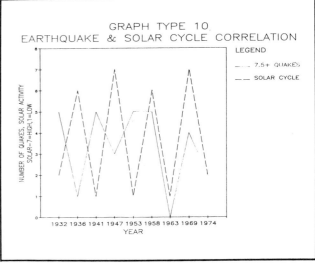

Fig. 7-13. Graph—combined solar and 7.5 earthquake activity—1932-1974.

**Computer
Note:**

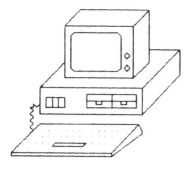

Many data base and word processing programs allow you to combine several files into one, making it easier to create smooth or combined data. Be sure, however, to have backup versions of your raw data before beginning work on the smooth data.

The smooth data should also include all the averages, totals, percentages, or other calculations that are necessary to tabulate and correlate your results. Depending on the nature of your experiment, you might require some more involved statistical or mathematical formulas.

If you haven't taken many math or science courses, you may need to consult with your math teacher to help select formulas that will appropriately analyze your raw data. When working with statistics, he or she may also provide guidance to determine how much variance, or which mathematical results, will prove or disprove your hypothesis.

In trying to correlate solar activity with earthquakes, it was insufficient to simply look at the graphs. A statistical test was necessary and the Correlation Coefficient Statistics Test, was used. Based on the data in Table 7-3, the statistical correlation was calculated, as shown in Tables 7-4 and 7-5.

**Computer
Note:**

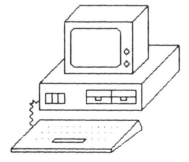

Your computer can also be very useful to you here. If there's a spreadsheet program available, you can use it to either calculate the results of an equation you've developed or to use one of its own built-in functions, such as count, sum, average, minimum, maximum, or standard deviations. If you've studied programming, you can also write your own programs to compute the statistical results.

Smooth data can also be graphed, by drawing a separate graph for a specified time period while combining the data for the variables and controls, as shown in Figs. 7-3 and 7-4. This, however, will result in a great many graphs if the experiment extended over a long period of time.

At the bottom of each smooth data table or graph, write one or two short paragraphs that summarize the data and explain briefly what the facts and numbers show.

ANALYZED DATA

The last element, analyzed data, will actually show the results, comparing the experimental groups with the control groups. The most effective method of presenting the analyzed data in a clear, accurate, and visually appealing way is to use charts and graphs. Line and bar graphs are the most common types of graphs used to present science project results as shown in Fig. 7-14.

Table 7-3. Raw Data—
Earthquakes (6.0 and 7.5) and Solar Activity.

Year	6.0 + Quakes	7.5 + Quakes	Solar Activity 1 = LOW, 7 = HIGH
1932	160	5	2
1933	175	3	3
1934	180	5	4
1935	135	5	5
1936	93	1	6
1937	97	4	5
1938	105	8	4
1939	96	5	3
1940	124	3	2
1941	132	5	1
1942	64	6	2
1943	93	11	3
1944	84	5	4
1945	64	2	5
1946	85	9	6
1947	49	3	7
1948	48	4	6
1949	94	6	5
1950	145	8	4
1951	125	2	3
1952	146	5	2
1953	117	5	1
1954	136	0	2
1955	132	0	3
1956	152	4	4
1957	183	11	5
1958	154	5	6
1959	148	3	5
1960	181	2	4
1961	156	3	3
1962	72	0	2
1963	63	0	1
1964	107	1	2
1965	167	3	3
1966	87	2	4
1967	104	1	5
1968	90	3	6
1969	118	4	7
1970	138	5	6
1971	133	7	5
1972	125	3	4
1973	108	7	3
1974	113	2	2

Regardless of whether you've used graphs in your raw and smooth data sections, here is where they can be used to their best advantage to show the continuity of the results. The objective is to combine as much as possible on one graph to show the correlation of data without a loss of clarity. Consider the use of color, if possible, to distinguish among the various elements. If not, design can be used, as shown in Fig. 7-15.

Pie charts, a very attractive way of diagramming results, will only lend themselves to projects that measure the division of a predetermined and specified quantity, for example, showing the percentage of a group of specimens. If your experiment uses that type of data, pie charts are appropriate and effective.

If you like working with graphs and your data lends itself to this method of presenting results, you might combine several graphing techniques. Remember, however, that as attractive as they are, graphs can only enhance the results; they will not compensate for deficiencies in research or experimentation.

Computer Note:

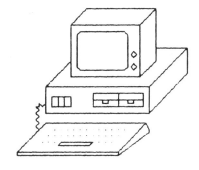

There are many computer graphics software packages on the market. Some can work on a dot-matrix printer as well as on an elaborate 12-pen plotter. They're a boon for anyone who wishes to include graphs and lacks either the talent or the patience to draw them accurately. Again, investigate using the school's computer if you have a great deal of graphing to do.

If you've used a spreadsheet to perform some of your computations, check to see if the software includes its own graphics package. If so, it will save you the work of recopying your results in order to graph them.

Be sure to title each graph or chart. Describe the data being analyzed and compared. Clearly and accurately label each axis, column, or row, including the unit of measurement used. For illustration, review the graphs shown earlier in this chapter.

The results section can also include any photographs you've taken to show what happened during your experiment. This can be very effective with experiments that have been conducted over several weeks, whose subject is easily and clearly photographed. For example, a project using plants, where the changes occur slowly over a period

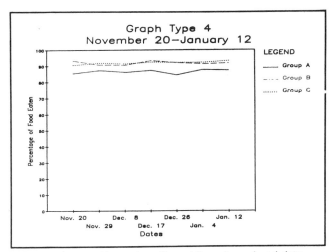

Fig. 7-14. Graph—*Mouse Appetite and Sound*—analyzed data.

of time, would be enhanced by using a series of photos.

Although sketches can add a great deal to the appearance of your project notebook, they are not considered as factual as a photograph. However, if you're talented in that area, by all means, add drawings to liven up your project when you start working on the final notebook or the project display.

Finally, in addition to charts, tables, graphs, and illustrations, you'll have to provide a written description of the results of the experiment. This should contain a narrative summary of observations and measurements. Describe any statistical work you've done, including any correlations between the experimental and control groups. Don't forge ahead and draw your conclusions, give "just the facts." In other words, this short statement should clearly and simply explain what happened and what the results showed.

CONCLUSIONS

At this point, you have done your experiment and organized your results. As you get ready to formulate the

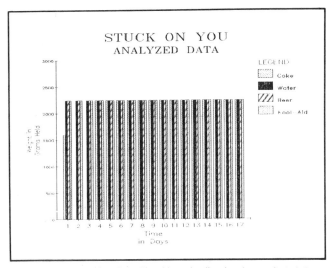

Fig. 7-15. Graph—*Stuck on You* (dental adhesives)—analyzed data.

		Table 7-4. Correlating 6.0 Earthquakes and Solar Activity Using Correlation Coefficient.			
		Y	X		
		6.0 + Quakes vs. Solar Activity			
Year	X_1	Y_1	X_1^2	Y_1^2	$X_1 Y_1$
1932	− 1.837	41.907	3.374	1756.167	− 076.983
1933	− 0.837	56.907	0.701	3238.407	− 047.631
1934	0.163	61.907	0.026	3832.477	0.10.091
1935	1.163	16.907	1.352	0285.847	019.663
1936	2.263	− 25.093	4.678	0629.658	− 054.276
1937	1.163	− 21.093	1.352	0444.915	− 024.531
1938	0.163	− 13.093	0.026	0171.426	− 002.134
1939	− 0.837	− 22.093	0.701	0488.101	018.492
1940	− 1.837	05.907	3.374	0034.892	− 010.851
1941	− 2.837	13.907	8.048	0193.404	− 039.454
1942	− 1.837	− 54.093	3.374	2926.053	099.368
1943	− 0.837	− 25.093	0.701	0629.658	021.003
1944	0.163	− 34.093	0.026	1162.332	− 005.560
1945	1.163	− 54.093	1.352	2926.053	− 062.901
1946	2.163	− 33.093	4.678	1095.146	− 071.580
1947	3.163	− 69.093	10.004	4773.843	− 218.541
1948	2.163	− 70.093	4.678	4913.028	− 151.611
1949	1.163	− 24.093	1.352	0580.472	− 028.020
1950	0.163	26.907	0.026	0723.986	004.386
1951	− 0.837	06.907	0.701	0047.706	− 005.781
1952	− 1.837	27.907	3.374	0778.801	− 051.265
1953	− 2.837	− 01.093	8.048	0001.195	003.101
1954	− 1.837	17.907	3.374	0320.661	− 032.895
1955	− 0.837	13.907	0.701	0193.404	− 011.640
1956	0.163	33.907	0.026	1149.684	005.526
1957	1.163	64.907	1.352	4212.918	075.486
1958	2.163	35.907	4.678	1289.313	077.667
1959	1.163	29.907	1.352	0894.428	034.781
1960	0.163	62.907	0.026	3957.291	010.253
1961	− 0.837	37.907	0.701	1436.941	− 031.728
1962	− 1.837	− 46.903	3.374	2124.564	084.672
1963	− 2.837	− 55.903	8.048	3035.238	156.298
1964	− 1.837	− 11.093	3.374	0123.054	020.377
1965	− 0.837	48.907	0.701	2391.894	− 040.935
1966	0.163	− 31.093	0.026	0966.774	− 005.068
1967	1.163	− 14.093	1.352	0198.613	− 016.390
1968	2.163	− 28.093	4.678	0789.216	− 060.765
1969	3.163	− 00.093	10.004	0000.008	− 000.294
1970	2.163	19.907	4.678	0396.288	043.058
1971	1.163	14.907	1.352	0222.218	017.336
1972	0.163	06.907	0.026	0047.706	001.125
1973	− 0.837	− 10.093	0.701	0101.868	008.447
1974	− 1.837	− 05.093	3.374	0025.939	009.355

E = SIGMA = the sum of . . .

$$0.143 = \frac{Ex_1 Y_1}{\sqrt{(Ex_1^2)(Ey_1^2)}}$$

"No Correlation, Statistically"

conclusions, you may wonder, "how do conclusions differ from results?"

Your results, as we've just discussed, stated what happened in your experiment, including any necessary mathematical or statistical interpretations and data correlation. You've also created charts, tables, and graphs to illustrate the data, and make it easy to see and interpret.

The word "interpret" brings us to the last scientific step in your project; the conclusions. Since the beginning of your project, as you researched the background material, and executed your experiment, the purpose was to answer a question and to prove the hypothesis. When you draw your conclusions, you will be analyzing your results to determine what you've accomplished, in other words, what was really learned from the trials and testing (or trials and tribulations, if you prefer!) of your experiment.

	Y	X			
Table 7-5. Correlating 7.5 Earthquakes and Solar Activity Using Correlation Coefficient.					
	7.5 + Quakes vs. Solar Activity				
Year	X_1	Y_1	X_1^2	Y_1^2	$X_1 Y_1$
1932	−1.837	0.907	3.374	00.823	−01.666
1933	−0.837	−1.093	0.701	01.194	00.915
1934	0.163	0.907	0.026	00.823	00.148
1935	1.163	0.907	1.352	00.823	01.055
1936	2.263	−3.093	4.678	09.576	−06.690
1937	1.163	0.093	1.352	00.008	−00.108
1938	0.163	−3.907	0.026	15.265	00.637
1939	−0.837	0.907	0.701	00.823	−00.759
1940	−1.837	−1.093	3.374	01.194	02.008
1941	−2.837	0.907	8.048	00.823	−02.573
1942	−1.837	1.907	3.374	03.637	−03.503
1943	−0.837	6.907	0.701	47.707	−05.781
1944	0.163	0.907	0.026	00.823	00.147
1945	1.163	−2.903	1.352	04.381	−02.434
1946	2.163	4.907	4.678	24.079	10.614
1947	3.163	−1.093	10.004	01.194	−03.457
1948	2.163	−0.903	4.678	00.008	−00.201
1949	1.163	1.907	1.352	03.637	02.217
1950	0.163	3.907	0.026	15.265	00.636
1951	−0.837	−2.093	0.701	04.381	01.752
1952	−1.837	0.907	3.374	00.823	−01.666
1953	−2.837	0.907	8.048	00.823	−02.573
1954	−1.837	−4.093	3.374	16.753	07.519
1955	−0.837	−4.093	0.701	16.573	03.425
1956	0.163	−0.093	0.026	00.008	−00.015
1957	1.163	6.907	1.352	47.707	08.033
1958	2.163	0.907	4.678	00.823	01.962
1959	1.163	−1.093	1.352	01.194	−01.271
1960	0.163	−2.093	0.026	04.781	−00.341
1961	−0.837	−1.093	0.701	01.194	00.915
1962	−1.837	−4.093	3.374	16.753	07.519
1963	−2.837	−4.093	8.048	16.753	11.612
1964	−1.837	−3.093	3.374	09.576	05.682
1965	−0.837	−1.093	0.701	01.194	00.915
1966	0.163	−2.093	0.026	04.381	−00.341
1967	1.163	−3.093	1.352	09.576	−03.597
1968	2.163	−1.093	4.678	01.194	−02.364
1969	3.163	−0.093	10.004	00.008	−00.294
1970	2.163	0.097	4.678	00.823	01.962
1971	1.163	2.907	1.352	08.451	03.381
1972	0.163	−1.093	0.026	01.194	−00.178
1973	−0.837	2.907	0.701	08.451	−02.433
1974	−1.837	−2.093	3.374	04.381	03.845

E = SIGMA = the sum of. . .

$$0.182 = \frac{E x_1 Y_1}{\sqrt{(E x_1^2)\, E y_1^2}}$$

"No Correlation, Statistically"

The project conclusions should be stated and explained in a paper of approximately three pages. In it, you'll analyze the results in light of your original assumptions, giving an interpretation of the data. You'll also critique your own project techniques, including the experimental design itself, and suggest future improvements on it.

Begin your conclusion paper by restating your question or hypothesis. Next, compare the results to your original hypothesis. The result data may concretely establish your theory to be true or false. However, it is also possible that the results were inconclusive, which means that although there may be a trend in your data, it is not strong enough to prove or disprove your hypothesis.

In formulating conclusions, patterns are what you're looking for. Closely examine your tables, graphs, and charts to see if a trend clearly emerges. The most important thing is to review your results critically and *without bias* in order to reach a definitive conclusion.

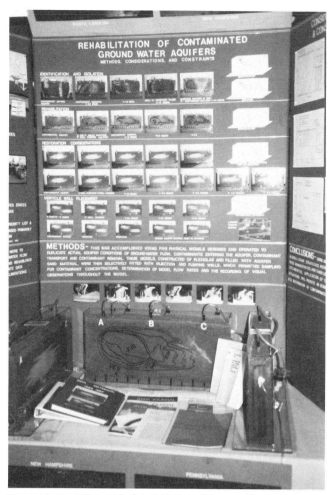

Fig. 7-16. Terri Newport; *Rehabilitation of Ground Water Aquifers.*

correlation at all.

Another possible reason for inconclusive results is a weakness in the experimental design itself. In the course of a science project lasting several weeks, which is conducted in a home environment rather than a fully equipped lab, it is sometimes impractical to perform a sufficient number of tests on enough specimens to overwhelmingly prove something.

The experiment graphed in Fig. 7-15 hypothesized that an orthodontic adhesive would weaken in various substances that were high in sugar. With one exception, the bonds did not break under pressure. The participant concluded that although the theory was disproved, it may have been because the trials were not carried on long enough, or the pull to loosen the bond was too weak.

In writing your conclusions, use your research as well as your results to explain the conclusions reached. The clearest and easiest way to do this is to discuss each fact or occurrence in a separate paragraph, referring to the experimentation and analysis that was done to reach the conclusion. Finally, write a summary that restates the hypothesis and conclusion as supported by the results.

If the results show a conclusive direction, you can happily state either that your assumption was correct and you've proven your hypothesis, or that the experiment failed to demonstrate that your theory was true. In any event, explain why your results occurred. This is important even if the hypothesis was correct.

For the project correlating mouse appetite and sound, the experiment proved the hypothesis, as shown by the consistently greater amounts and percentages of food eaten by the subjects in the experimental groups. (Refer to Fig. 7-14.)

You'll also want to explain results that disprove the hypothesis. The most likely reason is that the theory was incorrect. That's fine—a hypothesis is simply an educated guess, and refuting it is as scientifically valid as conclusive proof. In the earthquake and solar correlation project, although the graphs seemed to indicate a relationship, as shown in Figs. 7-7 and 7-8, the Correlation Coefficient Statistical Test did not. The statistical test required $+.82$ or $-.82$ to indicate a relationship. Because the calculated result, as shown in Table 7-4, was 0.143, and in Table 7-5, the calculated result was 0.182, there was no statistical

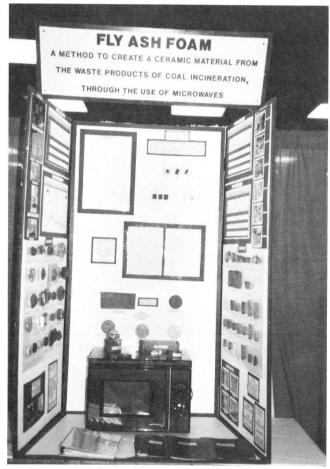

Fig. 7-17. Caroline K. Horton; *Fly Ash Foam: A Method to Create a Ceramic Material from the Waste Products of Coal Incineration through the Use of Microwaves.*

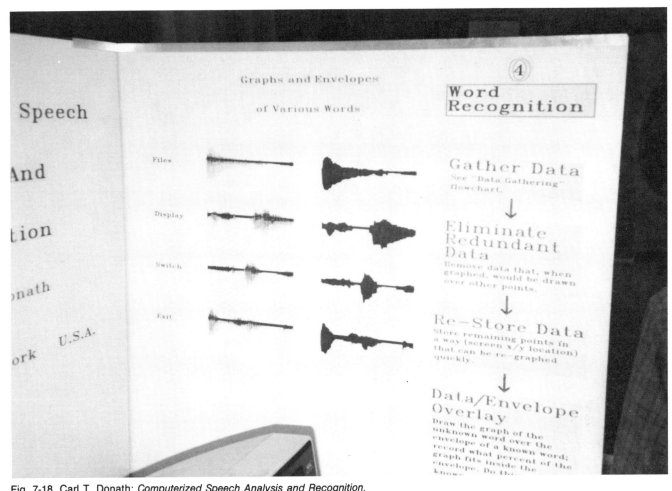

Fig. 7-18. Carl T. Donath; *Computerized Speech Analysis and Recognition.*

IMPROVEMENTS AND ENHANCEMENTS

Regardless of whether the project achieved its objectives, you'll need to analyze your experiment. *Realistically* reexamine the steps of your procedure in light of your results. Discuss the strengths and weaknesses of the design. Be honest—there are always some of each! Finally, suggest future modifications or improvements.

If this project has inspired you to continue research and experimentation in this field, briefly state how you plan to proceed. Even if you do not intend to pursue the subject, perhaps because it is beyond your resources or abilities, speculate on further work that could be done. However, keep in mind that several of the big winners at ISEF had been working on variations of the same project for many years, refining and developing their work, building each year on the strengths and weaknesses of their prior year's project.

Terri Newport of Stratford OK entered the 36th ISEF with a project entitled *Ground Water Pollution— Identification, Isolation and Rehabilitation*. The project did not win any awards, but the following year, she worked in the same general category. She expanded and enhanced the scope of her project and built actual models for her 37th ISEF entry *Rehabilitation of Contaminated Ground-Water Aquifers: Methods, Considerations and Constraints*. This project won her six awards, including the coveted four year scholarship to Ohio State University.

"REAL LIFE" APPLICATIONS

A powerful ending to your conclusions paper is a discussion of any potential practical value that your experiment may have. This will show teachers and judges not only your analytical ability as it relates to scientific reasoning, but a well-rounded approach, relating your work to other fields of endeavor. Looking at those types of relationships have brought a great deal of personal satisfaction to many students. In some instances, there is even the potential for professional recognition and commercial reward.

Caroline K. Horton's engineering project *Fly Ash Foam: A Method to Create a Ceramic Material from the Waste Product of Coal Incineration through the Use of Microwaves* produced a product whose qualities exceeded that produced at the Los Alamos National Laboratories, where she was conducting her research. She is now applying for a patent for her procedure.

Fig. 7-19. Scott Smith; *Computer System for the Handicapped.*

On the other hand, both Carl Donath, with his *Computerized Speech Analysis and Recognition* and Scott Smith, with his *Computer System for the Handicapped*, have received a less tangible, but very real reward.

They can take particular satisfaction in the possible contribution their projects can make towards using the growing computer technology to make a better life for others.

Chapter 8

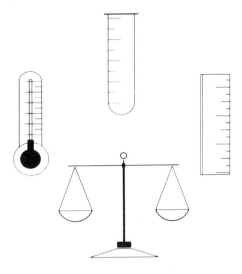

Best Foot Forward!

"Clothes don't make the man," says the old expression, but they can certainly help dress him up! Now that you've done the research, completed your experiment, analyzed and graphed the results and formulated your conclusions, you're ready to present your project.

As with any other presentation, it should be complete and accurate, as well as something that will generate immediate interest in your work. Although these might seem like opposing goals, they can both be satisfied, because the project presentation consists of two parts, a science fair notebook, and the display, or backboard.

THE SCIENCE FAIR NOTEBOOK

The Science Fair Notebook will contain all written and visual materials associated with your project. If this project was required, there's some good news and bad news. The bad news is that the notebook will probably constitute the lion's share of your grade, but the good news is that almost all the work is already done. Now, it's simply a matter of getting everything into the most attractive and presentable form.

You will want to put your best foot forward in this notebook, so start out by using the appropriate cover. Your teacher may have his or her own specific requirements for the size and type of folder, but if not, a good guideline is to use 3-hole, 8 1/2 × 11 paper and folder.

Use regular bond paper, since erasable bond smudges, and thin, onion skin paper may cause the reader to see double. *Always* double-space all written work, with the possible exception of tables, bibliography entries, or footnotes written in standard form, as shown in Chapter 4.

Some teachers may allow handwritten notebooks. If so, use ink only (no erasables please—they smudge!). The only possible exception is the log section, which represents raw data. However, type or use computer printouts wherever possible. The easier your work is on the teacher's eyes, the happier he or she will be. And a happy teacher usually gives better grades!

Now, let's review the items to be included in your science fair notebook.

1. Required Forms

The forms include the entry form, with the project description shown in Fig. 8-1.

If you used live vertebrates or tissue samples in your experiment, include the required forms (refer to Figs. 6-5 and 6-6). If you used human specimens, include the forms, as shown below.

2. Title Page

3. Table of Contents

Remember that although the Table of Contents goes right in the beginning of the notebook, it will probably be the last thing you'll create, once the pages are numbered. An example of a good format is shown in Fig. 8-4.

4. Abstract

The abstract is a shortened, or abbreviated, version of your background research paper. It should be about 200 words, and should summarize the hypothesis, the procedure, and the results. Keep in mind that you may need to fit the abstract, or some shortened version of it, on your entry form. Refer back to the project summary in Fig. 8-1.

GREATER SAN DIEGO SCIENCE AND ENGINEERING FAIR (GSDSEF)

CERTIFICATE OF ENTRANCE
(TYPE OR PRINT CLEARLY)

OFFICIAL
USE
ONLY

NAME OF ENTRANT (LAST, FIRST, MIDDLE)

HOME ADDRESS

GRADE _____ FEMALE ☐

AGE _____ MALE ☐

PROJECT ADVISOR (First Initial, Last Name)

CITY, STATE, ZIP

SCHOOL   CITY

ENTRANT'S SOCIAL SECURITY #
(Optional — Helpful to Awards Staff)

HOME PHONE

HAVE YOU EVER HAD A PROJECT ACCEPTED
BY THE GSDSEF **BEFORE** THIS YEAR?   NO ☐   YES ☐

IF YES, HOW MANY YEARS? _____

PARENT(S) / GUARDIAN(S) NAME(S)

1. Biology / Microbiology	7. Chemistry / Biochemistry
2. Botany	8. Earth / Space Sciences
3. Zoology	9. Physics
4. Medical Sciences	10. Engineering / Electronics
5. Human Psychology / Social Sciences	11. Mathematics
6. Animal Behavior	12. Computers

WAIVER — ALL FAIR ACTIVITIES INCLUDING TOURS
IN CONSIDERATION OF YOUR PERMITTING THE UNDERSIGNED STUDENT TO TAKE PART IN
THE GREATER SAN DIEGO SCIENCE AND ENGINEERING FAIR (GSDSEF), WE WAIVE ALL CLAIMS
AGAINST THE GSDSEF, UNION / TRIBUNE PUBLISHING CO., THE COPLEY PRESS INC. AND THEIR
RESPECTIVE OFFICERS, DIRECTORS, GOVERNORS, COMMITTEES, EMPLOYEES AND AGENTS FOR
INJURY TO OR DEATH OF PERSONS OR LOSS OR DAMAGE OF PROPERTY IN ANY WAY
OCCURRING IN CONNECTION WITH THE GSDSEF AND WE AGREE TO INDEMNIFY AND HOLD
THEM HARMLESS AGAINST ALL SUCH LIABILITY
WE ALSO AGREE THAT THIS PROJECT WILL BE KEPT ON DISPLAY AT THE FAIR UNTIL 5:00
P.M. SUNDAY AND WILL BE REMOVED BY 6:00 P.M. SUNDAY.

Parent/Guardian's Signature _____

Date _____

DISPLAY AREA REQUESTED (CHECK ONE)
(MAXIMUM SIZE — 122cm x 76cm or 48 in. x 30 in.)
☐ TABLE
☐ FLOOR (NO TABLE)

ELECTRICAL OUTLETS REQUIRED
(110V 60 CYCLE ONLY)
NUMBER OF OUTLETS _____
AMPS _____ WATTS _____

DOES PROJECT INVOLVE EXPERIMENTS WITH LIVE
VERTEBRATE ANIMALS? _____
(LIVE VERTEBRATE ANIMALS MAY NOT BE DISPLAYED)

NOTE: IF LIVE VERTEBRATE ANIMALS ARE USED,
GSDSEF FORM 2 MUST BE SUBMITTED WITH
WITH THIS CERTIFICATE

DOES PROJECT INVOLVE EXPERIMENTS WITH
HUMAN SUBJECTS? _____

NOTE: IF HUMAN SUBJECTS ARE USED
GSDSEF FORM 3 MUST BE SUBMITTED
WITH THIS CERTIFICATE

DOES PROJECT INVOLVE EXPERIMENTS WITH
TISSUE SAMPLES OR PARTS OF HUMAN
OR VERTEBRATE ANIMALS? _____

NOTE: IF TISSUE SAMPLES / PARTS ARE USED,
GSDSEF FORM 4 MUST BE SUBMITTED
WITH THIS CERTIFICATE

CERTIFICATION OF COMPLETENESS AND ACCURACY OF INFORMATION AND COMPLIANCE WITH ALL REGULATIONS

SIGNATURE OF ENTRANT

SIGNATURE OF SCHOOL PROJECT ADVISOR

PROJECT TITLE (10 OR FEWER WORDS)

PROJECT SUMMARY — (200 - 250 WORDS DESCRIBING PROBLEM, METHODS AND SIGNIFICANT FINDINGS, NEATNESS AND THOROUGHNESS IMPORTANT.)

TYPE SINGLE SPACE TO DOTTED MARGIN

GSDSEF FORM 5 (1987)

LIST **ALL** PROFESSIONAL ORGANIZATIONS, SCIENTISTS AND OTHERS, (IF PARENTS, NEIGHBORS — NO NAMES PLEASE)
FROM WHOM YOU WERE ABLE TO OBTAIN ADVICE AND / OR HELP IN DOING YOUR PROJECT.

Fig. 8-1. Project entry form.

CERTIFICATION OF COMPLIANCE OF RESEARCH INVOLVING HUMAN SUBJECTS

Name of Entrant _____ School _____

Project Title _____

Because federal regulations have become increasingly more rigid, students must plan carefully before undertaking research which involves the use of human subjects in either behavioral or biomedical studies. This will protect subjects from unnecessary exposure to physical or psychological risks and experimenters and schools from legal complications.

A **human subject** is legally defined as:

A person about whom an investigator (professional or student) conducting scientific research obtains (1) data through intervention or interaction with the person or (2) identifiable private information.

A **subject at risk** is legally defined as:

Any individual who may be exposed to the possibility of injury, including physical, psychological or social injury, as a consequence of participation as a subject in any research . . .

Students using human subjects must comply with all regulations that reflect the will of society and plan proper methodology for the protection of those subjects. It is essential that they be alert to humane concerns at all times.

The following steps must be taken **before** any student begins research involving human subjects:

1) The student completes the "Research Plan" section of this form (GSDSEF-3) and submits it to the sponsoring teacher.

2) The sponsoring teacher reviews the "Research Plan" and determines if **ANY POTENTIAL** physical, psychological or social risk is involved (as defined in **subject at risk** above).

    a) If none is apparent, the teacher signs the certification. (No additional certification is necessary.)

    b) If any question exists, the student must redesign the experimental study or plan a different study.

        NOTE: If assistance is needed to evaluate a particular research plan, call the appropriate screening committee representative (as listed in the GSDSEF Student Guide.)

NOTE: Any project involving human subjects that is developed with the advice and assistance of personnel at a medical/scientific organization must comply with any regulations of that organization requiring approval of its Institutional Review Board and Informed Consent Certification.

RESEARCH PLAN

Describe proposed experimental procedures:

Explain why human subjects are proposed for this experimentation:

Fig. 8-2. Certification of compliance of research involving human subjects.

Describe and assess any potential risk (Physical, psychological, social, legal or other):

Describe the potential benefits to the individual or society:

Signature of student _____ Date _____

CERTIFICATION

CERTIFICATION BY TEACHER/ADVISOR of compliance with federal regulations for the protection of human subjects in behavioral and biomedical research. (**Must** be completed **before** the start of experimentation.)

I certify that, upon reviewing this research plan, I found that the experimental procedures constitute no physical, social or psychological risk to either experimenter or subjects.

I agree to supervise this experimentation and will insure that it is conducted in a humane, risk-free manner.

Signature _____ Name (type or print) _____ Date _____

Institution _____ Title _____

Institution Address _____ Phone _____

Home Address _____ Home Phone _____

This form, properly completed, must be part of the carefully planned procedures for any experiment involving human subjects. It must accompany any such project exhibited at or presented for any public display associated with the Greater San Diego Science and Engineering Fair.

Form GSDSEF-3 (1987)

HOW MUCH DO WE REALLY KNOW ABOUT SEX?
By:
Erin Filner

Crawford High School
P. Wright
February 28, 1986

Fig. 8-3. Title page.

5. Background Research

The Background Research Paper is the same research paper you finished at the beginning of the project. However, you probably had corrections marked on it by your English and Science teachers. Incorporate their suggestions now, in order to fix, clarify, and edit your work.

You might have done some additional research as you progressed with the experiment, especially if you got some professional advice and assistance. Incorporate this information into the final paper to make it more thorough and complete.

6. Bibliography

Just as with the background research paper, the bibliography should be almost complete. Make any corrections in format, spelling, or style that your teachers have requested. If you've added information to the research paper, or conducted interviews, etc., remember to credit any new or additional sources.

7. Statement of the Problem or Question

Restate the question or problem that expresses the purpose of the experiment.

8. The Hypothesis

What was the educated guess that you made, which explained what you tried to prove with the project? Check to see that you've identified the subject of the experiment that you separated into experimental and control groups. Make certain that you've named the dependent and independent variables, and the controls. If your teacher asked you to reword it when you handed it in the first time, now is the time to make these corrections.

Please be honest and state your *original* hypothesis even if the experiment disproved it. Remember that this information is just as valuable as a correct guess!

9. Procedures

List all your procedures, numbering them sequentially. Avoid the first person, use the past tense and include diagrams, where necessary. If you altered the procedures in any way since the time they were originally approved, be sure to modify them now, before including the list in your final notebook.

10. Materials List

Include anything and everything you bought, begged, borrowed, or built.

11. Variables and Controls

Fully describe each variable and control. Explain its role in the experiment and show how it was managed and/or monitored.

12. Results

Include raw, smooth, and analyzed data, in all forms, including all the charts, graphs, tables, photographs, and diagrams you've created or collected in the course of the experiment. Be sure that all results, in whatever format, are neat and legible. Accurately and clearly label and title all material. Include the narratives that describe your observations and the outcome of the experiment, as shown in Chapter 7.

13. Conclusions

Since you've just finished the conclusion paper, all you'll need to do is make any necessary corrections and type or print the final copy.

14. Acknowledgments

Here, you get the opportunity to thank everyone who has helped you to complete your project. It is a good idea not to identify teachers, school, or family by name, because at certain levels of competition, you are required to remain anonymous. See Fig. 8-5 for an example of a well-written, anonymous acknowledgement section.

TABLE OF CONTENTS

Section	*Page*
Background	2
Bibliography	10
Problem	11
Hypothesis	12
Procedures	
Description	13
Photographs	15
Variables/Controls	17
Materials	18
Results	20
Tables	22
Graphs	30
Photographs	45
Conclusions	50
Acknowledgements	56
Log	57

Fig. 8-4. Table of contents.

ACKNOWLEDGEMENTS

I would like to thank, for helping me with my science project this year, my mother, my father, and my brother.

I would also like to thank   of the San Diego State University Natural Sciences department, for helping me find the information that I needed to do my science project.

I would further like to thank the people at the San Diego branch of the National Oceanographic and Atmospheric Administration at the Southwest Fisheries Center and the librarians at the UCSD Scripps and SDSU libraries for helping me with the search for information within their library.

Last and definitely not least, the people I could not possibly forget for giving me all the support and help I needed: , my science teacher, and   , my algebra teacher, for helping me with the statistics correlation test.

Fig. 8-5. Acknowledgments.

15. Project Log

The Project Log is one place where the "neatness counts" rule doesn't necessarily apply. Include your diary, working log, rough notes, and drafts of tables and graphs.

When you've reprinted and collected everything, review all your materials at least once, preferably with someone helping you. Make sure that you have all the sections in order, and that the pages within each section are in the proper sequence.

Number the pages and create the table of contents. If you're going to number your pages by hand, have someone help you by writing the numbers on the page, if your handwriting isn't especially legible. Another alternative is to insert each page into the typewriter to put the numbers either on the top or bottom of each page.

Computer Note:

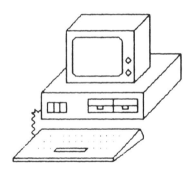

If you have all the information on diskettes, obviously you only need to make corrections and reprint, taking special care to have proper margins and format. At this time, consider using a fresh ribbon in your printer.

Page numbering could also be done using your word processor or other software. Consult the user manuals to learn how to include different files in one document. If your notebook materials come from various sources, be sure to allow for inserts when deciding to let the program number the pages.

Now that you're done with your notebook, let me repeat that a neat, well-presented notebook will certainly enhance a well-done project, but it cannot compensate for a poorly researched, inadequate experiment.

THE SCIENCE FAIR DISPLAY

The final step is to construct your project exhibit. This presentation is not your product, the research and experiment are. What this represents is your advertisement. A good display will attract attention to your work, inviting judges to look at your notebook and investigate your project more thoroughly.

Just as with the notebook, the most attractive, elaborate backboard in the world won't substitute for mediocre work. However, because you've followed this step-by-step guide, your project will be excellent, deserving a first-class exhibit.

The purpose of the display is to summarize your project. Read that sentence again and remember the key word, *summarize*. Do not try to recreate your entire project notebook. Instead, simply cover the main points and the highlights.

Most science project displays consist of three-sectioned, free standing backboards. The sections are normally hinged together, for easier folding and transporting to the various science fairs where you'll be competing. However, backboards with as few as two sections, or as many as four, have been used successfully.

When looking at fairs at all levels, we've seen backboards made of a variety of elements. The best are constructed of rigid, durable, fireproof goods, such as masonite, pegboard, plywood or plexiglas. Some science fairs may prohibit displays made of cardboard or other paper products, due to fire regulations. Although the actual construction may require some time and expense, the basic backboard can later be stripped of the display materials and used for subsequent science projects.

The normal height for a science fair display is between 3 and 5 feet. When deciding, first check to see if the science fair itself has restrictions on overall size. These limits are usually imposed due to the size of tables available. Occasionally, there is a display that is so large that it cannot be placed on a table. This happens most often with engineering projects. Be sure to check with those in charge of your local science fair before designing such a presentation.

Next, decide whether or not you plan to cover the basic backboard or keep it bare. If you're using pegboard, you may wish to use hooks to attach material. Some students

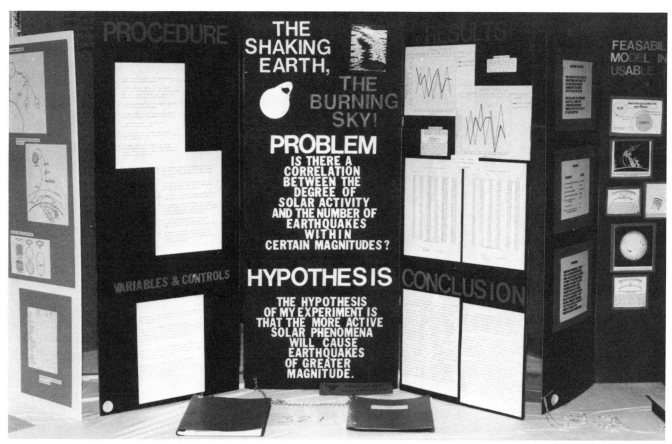

Fig. 8-6. Mike Iritz; *The Shaking Earth, The Burning Sky.*

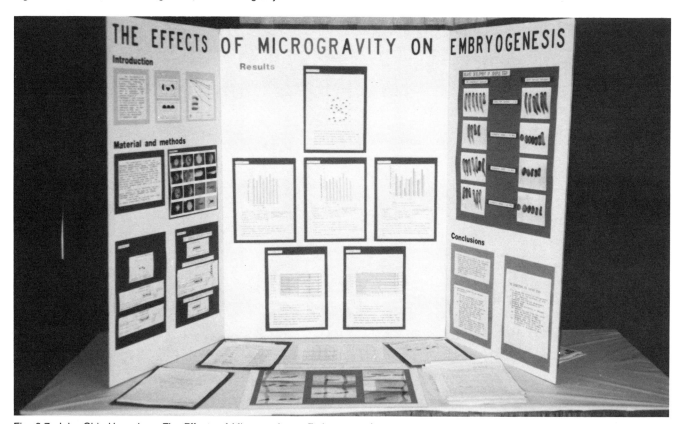

Fig. 8-7. John Chin-Hung Leu; *The Effects of Microgravity on Embryogenesis.*

Fig. 8-8. Francis Vazquez; *Comparative Study of Raw and Pasteurized Milk from Different Dairies of Puerto Rico.*

also created a panel to drape over the front of the display table, in which pockets for the Log Book, Abstract, and Research were included.

The best way to insure a well-designed backboard is to make a map or blueprint of your project display. Decide what to put on each of the three panels. Many have included the title, the question and the hypothesis on the center panel, reserving the sides for the procedures, results and conclusions. However, there's no hard and fast rule for this, as you can see from both the photographs and the sample backboard layouts included in Fig. 8-11 A and B.

Be sure to figure out the dimensions of everything to be included, as well as the size and style of the lettering you will use. One thing to keep in mind is that when displaying at a science fair, your notebook will be attached to the lower left portion of the middle section by a piece of chain, twine or other strong material. Be sure to leave a small spot free for that attachment.

This layout is strictly for your use, so feel free to create several versions, until you have one that is suitable and attractive. Once you've decided on a blueprint, show it to your teacher before you actually begin mounting things on the backboard.

have painted their backboards attractively, and others glued velcro strips to the basic backboard, in order to attach materials later. If you do plan to cover the panels, keep in mind that construction paper comes in 12-inch sheets, which will make it harder to cover an odd-sized backboard. Other covering materials may also come in standard sizes, which should be considered when choosing the size.

Your display *must* include summaries of the problem, hypothesis, procedures, results, and conclusions. You can do this using all written material, or you may include graphs, photographs, drawings, tables, or other artwork. Remember, however, that all this information must fit onto the backboard you have selected. However, some students have very creatively solved the problem of too much material and too little room on a backboard. Francis Vazquez created rows of overlapping flip charts to display his many graphs on the backboard for his project *A Comparative Study on the Chemical and Bacteriological Phases in the Processing of Raw and Pasteurized Milk from some Dairies in Puerto Rico.*

In the display for *Effect of pH Upon the Culture: Penicillum Notatum* Shawn Bray not only created a "mini-diorama" in front of the backboard to display the molds, but

Fig. 8-9. Shawn Bray; *Effect of pH upon the Culture—Penicillin Notatum.*

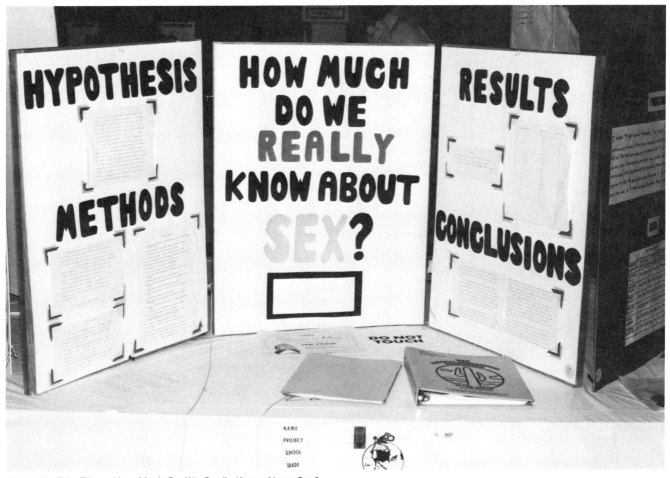

Fig. 8-10. Erin Filner; *How Much Do We Really Know About Sex?*

If you plan to display any equipment or specimens such as plants on the table in front of the backboard, plan to leave the bottom portion clear of text and illustrations, if necessary, because it may be blocked by these items. Notice that David Rodgers, when designing the display for his project *Growth Pattern of Wheat in Relation to Water Source* placed his written and visual material on the upper two-thirds of the backboard, leaving room for four trays of wheat, and tags for each.

In Caleb John's project *What Effect does Drafting have on Energy Requirements,* he stopped the written material about 3/4 of the way down the backboard to allow room to display the equipment.

Of course, if your material can be placed flat upon the display table, as was done in Kelly O'Neill's project *Which Wood Burns Best in a Fireplace,* you need not allocate any extra space for it.

Remember, however, that no live insects or vertebrates can be displayed at a science fair.

As you've noticed, at least some portions of your display will consist of written material. The problem and hypothesis, which are already short statements, can probably be used without any further editing. However, you can show display printed matter in several ways.

—Typewritten.
—Computer Printout. You can experiment with different sizes and fonts using your computer.
—Hand lettering.
—Stencilling or press-on letters.
—Lettering made on a Kroy or other labelling machine.

As shown in the sample display blueprints, be sure to use a title on everything you display, using larger lettering to make it stand out. When doing the written material, you will become an expert in the art of summarizing.

Some participants create backboards which consist mainly of written material, as in *The Study of the Wasp.* However, Akira Soya used illustrations and graphs with the text to create an interesting display.

The main thing to remember here is that "less is more." You'll want to be accurate using as few words as possible.

It's unnecessary to present all the facts on the display. Simply highlight the most important points. If you've done this well, the judges will take the time to examine the notebook and interview you to find the details. On the other hand, including too much information will make it look busy and difficult to read, perhaps discouraging a closer look.

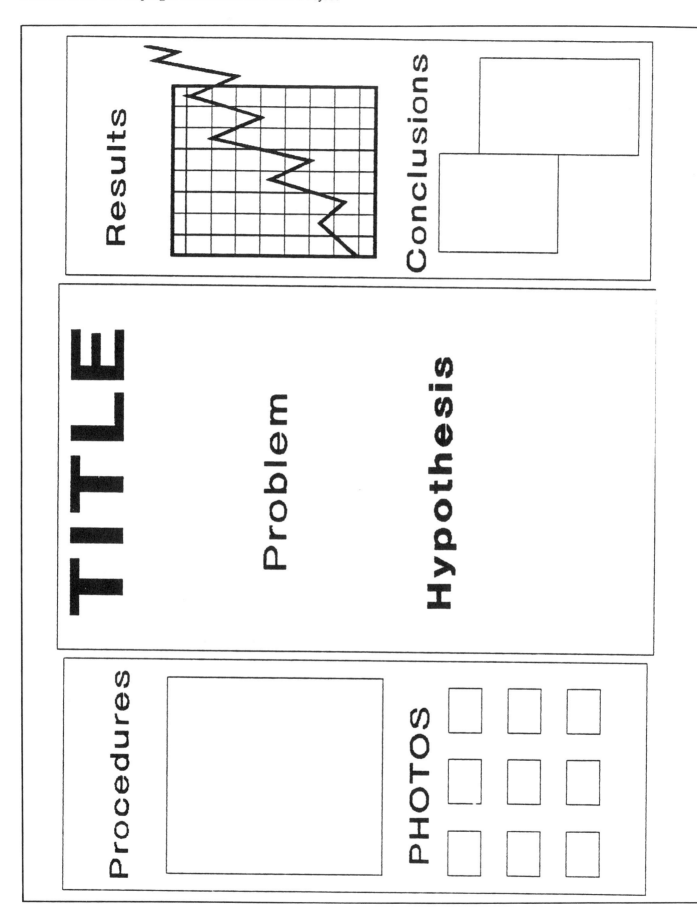

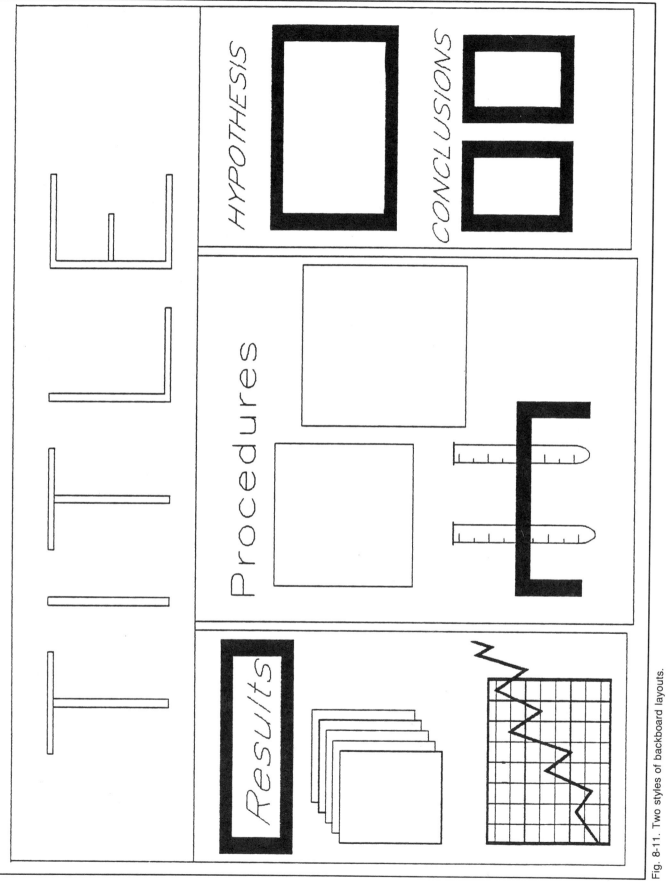

Fig. 8-11. Two styles of backboard layouts.

Fig. 8-12. David Rodgers; *Is There A Difference in the Growth Pattern of Wheat When Irrigated by Various Water Sources?*

If you decide to include graphs, charts, or tables on your backboard, it's a good idea to create enlarged versions, making them more visible and attractive. One year, we found an excellent solution to drawing endless lines with thick markers on construction paper, only to have them come out fuzzy. The art supply store had, in a final markdown basket near the door, rolls of glossy, colored tape in a variety of shades and widths. These can be used in either line or bar graphs for a professional and artistic look. Thin, black tape can be used to show gradations on graphs. Remember to make them accurate as well as attractive.

Photographs or illustrations can be mounted using hinges, corners, or glue. Be sure, that anything pasted or glued on your backboard will stay in place for the duration of the exhibit. If you, a friend, or relative are artistic, line drawings can also be an eye-catching addition.

A good cement useful for bonding paper is the best way to paste things on a backboard. Buy the proper adhesive for the material you are using, and follow the manufacturer's directions, for example, to ventilate the room properly when working with certain types of glue. Also investigate using staples, nails, push pins, or other methods of attaching material to your display.

At this point, it's also a good idea to select the colors for the display. Contrasting colors that attract attention without being loud or garish usually work well, and metallics are good attention getters.

Fig. 8-13. Caleb Johns; *What Effect Does Drafting Have on Energy Requirements?*

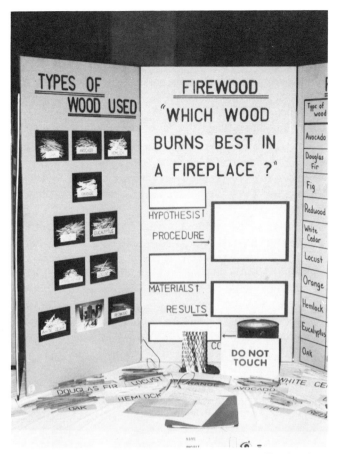

Fig. 8-14. Kelly O'Neill; *Which Wood Burns Best in a Fireplace?*

Fig. 8-15. Soya Akira; *A Study of the Wasp—The Mechanism of Determination of the Sexual Function in Vespula Lawisii.*

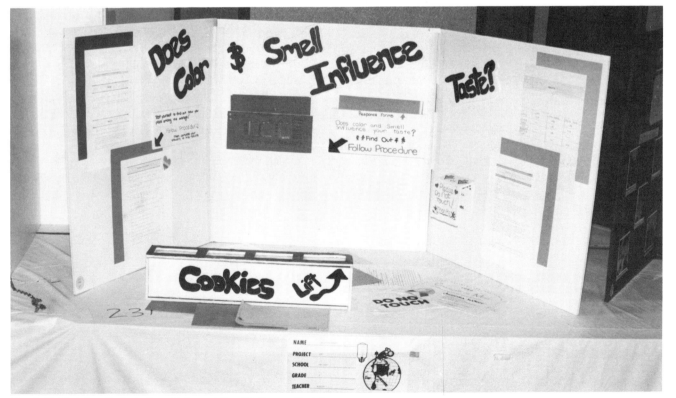

Fig. 8-16. Christina Hodnet; *Does Color and Smell Influence Taste?*

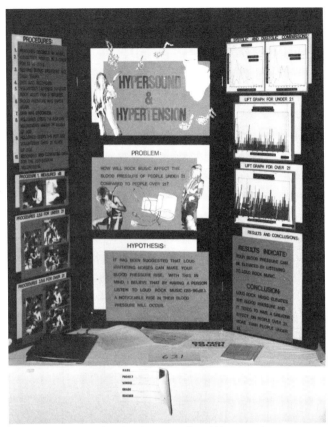

Fig. 8-17. Gregory Syrek; *Hypersound and Hypertension.*

Many students have also used colors that reflect the project itself, such as green for botany projects. In any event, if you're using several colors, check how they look together before making a final decision.

Always (no exceptions) have extra material on hand. A Sunday night, when the display is due first period on Monday morning, is the worst possible time to find that you've misspelled something on the very last sheet of construction paper. It's a mistake that has often reduced panicked families to rummaging through wastepaper baskets for an even halfway respectable scrap.

Careful planning and measuring, however, will reduce the need for such last-minute resourcefulness. Lettering seems to be the area most vulnerable to errors, so here are a few hints that are useful, regardless of whether you're using press-ons, stencils, or doing your own.

First of all, write out what you're going to say. Check the spelling of each and every word to be included. Decide on capitalization and punctuation, and check that for accuracy and completeness, too.

Next, check the number of lines the text will occupy. To do this, decide upon the size of lettering you'll use. Then, double check and make sure your message takes the same amount of room that you anticipated. Also verify that what you need to say will fit across the panel. If not, reach for a thesaurus to find alternate ways of stating the needed information.

If you're buying press-on letters, count how many of each letter, in each size, that you'll need for the entire display. Although this may seem like a tedious task, it may save money for buying extra sheets "just in case," or even worse, running out of E's at the last minute. If the worst happens, however, and you run out after the store has closed, remember that with a good eye, a steady hand, a sharp razor blade and fragments from unused J's and Z's, an F can become an E, a V can be transformed into an A, and an P can be made from an R!

When you're finally ready to do the lettering, to assure a professional looking job, take these steps:

1. Lightly draw lines across the sheet to be sure that your lettering is level.

2. Lightly mark the center of each line. Then, place the center letter(s) of your text on that spot and work outwards, until you've completed your line. If you're using stencils, you may wish to outline these in pencil and check the line before filling the forms with marker or paint. If, on the other hand, you're using press-ons, keep them *lightly* attached in case you need to remove them later.

Now that you're ready for the final steps, attaching written and visual materials, follow similar steps. Lightly mark exactly where the particular item is to go, and make sure it creates the effect you really want before gluing it into place.

Attach your notebook to the lower-left corner of the center panel and the display is done. Open up the backboard, step back and enjoy. You've done a great job!

Chapter 9

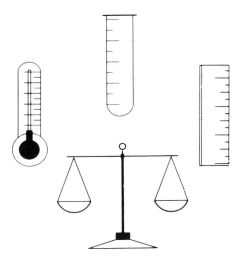

The Fair

The big day is approaching. You've done your best work, created an attractive presentation, and you're ready for your first science fair.

SETTING UP

On the day before judging, you'll need to set up your display in the exhibit hall. This area can be as simple as a school lunchroom or as elaborate as a civic center or convention hall. Regardless of where it's held, however, some things remain the same.

When you enter the exhibit area, your display will be checked for size, and then assigned a number, which will indicate your spot. Projects in the same category are always grouped together. This is for convenience, because there are often different judges for each category. Within a category, places on the exhibit tables are usually assigned by alphabetical order, so your location is the luck of the draw. The only exceptions are oversized displays or those needing electrical power.

Incidentally, project areas are usually guarded by professional security personnel. Don't worry, then, about your notebook or other display material getting "ripped off." Other exhibitors and their parents are also concerned about expensive computer equipment included as part of their display. Be assured that everything possible is done to protect your property.

Installing your project can be less hassled if you're prepared. If setting up involves crawling and climbing, wear old clothes. You may want to bring a small collection of tools, such as a hammer and nails, or a screwdriver, to fix anything that breaks in transport. A glue stick to fix corners of your display material that have become dog-eared might also be helpful. If you use light bulbs that can burn out or glassware

that can break, it's a good idea to have extras on hand for the duration of the science fair. Incidentally, having your own tool kit may make you the most popular person in the hall on set-up day.

JUDGING DAY

Understandably, you're nervous. You know your project is good, and your display is attractive but don't know what the judges are looking for. When you set up, you may have seen some other projects in your category. With a total lack of confidence, you may feel that everyone else's project is better than yours. Don't panic, though—opening night jitters are normal.

On judging day, most science fairs are closed to the public. Until the awards ceremony that afternoon or evening, it's just you, the officials and the judges. First, realize that the judges will be looking at many other projects on the same day that they're examining yours. One way of attracting positive attention, even *before* they review your work, is to appear organized and professional.

Here, I'm going to take a moment and sound a bit like your grandmother, but remember that her advice must have been good if it hung around for so long!

If possible, sleep well the night before. Fatigue and stress do not mix well when trying to make a good impression. Next, look well. You need not wear expensive clothes on judging day, but neat, conservative clothes (maybe the same ones you wore to Thanksgiving dinner) and a clean, well scrubbed look help create a positive image. Also act well. Don't eat, chew gum, clutch a soft drink, or slouch when the judges are walking through the exhibit area. Finally, speak well, both verbally and nonverbally. Smile and speak politely when questioned or addressed.

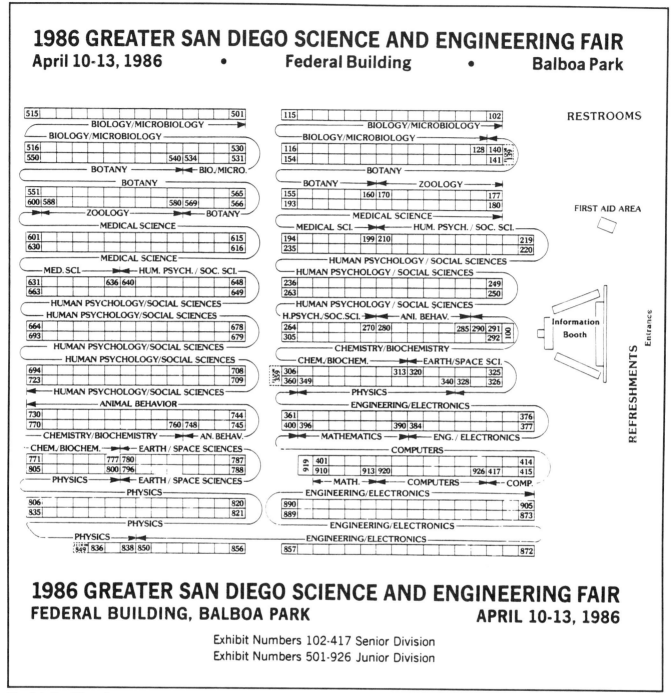

Fig. 9-1. San Diego Science Fair floor plan.

Although your project is being evaluated, not your manners or appearance, remember that the judges are only human. They'd rather spend time with an agreeable exhibitor than an unpleasant one.

It's also important to be ready to discuss your work. At some science fairs, each student is expected to give a short oral presentation about his or her project. At others, the judges will circulate through the exhibit area, looking at every project, and asking questions. When asked, don't wave your arm and say "it's all here on the backboard."

The judge already knows that. He or she wants to hear you express yourself.

Even if you're not expected to give a prepared oral presentation, it's a good idea to plan and rehearse what you will say. Check and practice the pronunciation of difficult words. Try to remember these helpful hints:

1. Throughout your discussion, remain friendly and polite. Maintain good eye contact, not staring at the floor or the ceiling. Stand up straight. (Hi, grandma!)

2. Be ready to introduce yourself and your project.

3. State your purpose and hypothesis clearly and distinctly, followed by a *brief* description of your procedures. If you have your specimens or apparatus on display, now is the time to point it out. Invite the judge to examine it more closely or try operating the device.

4. List your results. Now, point out any graphs, charts or tables, either in your notebook or on your backboard.

5. Discuss your conclusions, including whether or not the experiment proved your hypothesis.

At this point, even if you've given a prepared presentation, the judges will, no doubt, have some questions. Some common areas are:

—How or why did you get interested in the topic?
—Are there any aspects of the experiment or research

Fig. 9-2. Illustration of award ribbon.

that you might have changed or corrected, if you had the time?

—Do you intend to continue work in this area? If so, how? If not, why not?

—What practical applications or future use does your work have in "the real world?"

—Have you seen the article last month in the Bladedy-Blah Magazine by Dr. Such-and-such dealing with the further implications of etcetera and so forth?

First of all, don't be afraid to admit that you don't know an answer or that you haven't read the article or book in question. You'll make a far better impression with your honesty than with a futile attempt to "snow" an expert.

This brings us to another point. With either your answers or prepared speech, certainly try to be accurate and informative. However, don't try to give the judge all the information in your notebook in less than two minutes. Summarize, show and tell, covering the main points.

Science fair judges agree that the factors that come across most positively are knowledge and enthusiasm. A student who takes the path of least resistance by selecting a topic requiring a minimum amount of work is unlikely to rate high marks, even if he or she has a beautiful display. However, someone who has really worked to learn as much as possible and made an effort to follow good scientific procedures will impress the judges, even if the experiment has not worked out well. As we've all heard before, enthusiasm is contagious. Participants who are excited about their experiment and research show that they've gained the true benefit of doing a project and being in a science fair.

When the talk or the question and answer period is over, smile, shake hands and thank the judge. When he or she has moved on to the next backboard, you can breathe a sigh of relief. Now it's only a few more hours until you find out if you've won an award and the chance to advance to the next level.

AND THE WINNER IS . . .

That's right. You're the winner, before you even get to the awards ceremony. In fact, you've been a winner from the moment you started, as you selected your topic, completed your experiment, formulated your results and conclusions and created your display. Your perseverance in sticking with it through the difficult times, your flexibility in finding "alternate plan B" when your experiment seemed to be failing, and your creativity, curiosity, and talents have won you new knowledge and confidence, and perhaps even a life-long enthusiasm.

However, awards are nice! From the Oscars and Emmys to your local science fair, everyone wants the acclaim and recognition of their teachers, family and friends. There is usually a wide variety of awards. There are the awards of the particular science fair itself, which consist of first, sec-

ond, and third place in each category. Then, there are sweepstakes winners, who are judged to have the best overall projects. Finally, various private companies, research institutes, and military representatives give awards to projects that enhance the knowledge in their particular area. Such special awards are often conferred in such diverse areas as ophthalmology, nutrition, aerospace, and oceanography.

At the International Science and Engineering Fair, many such special awards are given. NASA, the International Aerospace Hall of Fame, the U.S. Armed Forces, corporations, such as General Motors and Eastman Kodak, and many others confer prizes. Trips to other science fairs, both in our country and abroad, are coveted prizes. There are also many scholarships awarded, among them, complete four-year scholarships to Ohio State. In the International Fair, the highest award, given to two projects selected from the Grand Prize winners in each category, is the Glenn T. Seaborg Nobel Prize Visit. This consists of an all-expense-paid trip to attend the Nobel Prize ceremonies in Stockholm, Sweden.

Fig. 9-3. Federal Building, Balboa Park, San Diego, California.

Fig. 9-4. Fun at the Fair!

If you've won an award, congratulations!! Hopefully, it's the first of many. But, if you didn't get a prize, don't let your disappointment spoil the science fair for you, or dampen your enthusiasm for future competition. If you're lucky enough to receive copies of your evaluations, you might get some good ideas on how to improve your project, and could possibly consider restructuring your project and entering it the following year. But, whatever you've come away with, you've gained a working knowledge of the scientific method and an insight into an area of science.

MEET ME AT THE FAIR

Once the excitement and tension of judging and awards night is over, enjoy yourself. Most science fairs go on for several days, to allow both school children and the public at large the opportunity to review the projects.

If you like meeting and talking with people, this might be fun, especially if you have the kind of project that generates many questions. Some exhibitors even prepare hand-outs describing their work to give out to interested viewers. Most students I've spoken to, especially first time participants, have spoken with pride of meeting an acquaintance who didn't even know they were in the fair.

You'll also get the chance to look at others' projects. For comparison, interest, or perhaps ideas on future projects, most participants enjoy looking at others work.

Finally, in many cities, museums, hospitals, universities,

Fig. 9-5. Struck by lightning!

and other institutions join together to congratulate and celebrate the participants in their local science fair. In San Diego, California, the fair is held in Balboa Park, the site of most of the city's museums. For the duration of the fair, an exhibitor badge gains the student free admission to any museum. There are also lectures and guided tours, especially for participants. The University observatory, a behind-the-scenes look at the zoo, or a trip to a medical center, can be fun and informative. Counseling sessions, given by professionals to advise students about various scientific careers, are also available.

You'll be required to keep your project on exhibit until the specified end of the fair, or suffer the consequences. This is to insure each student will check out his or her own project materials when the fair is over.

In the end, at 5 P.M. on Sunday, April 13, 1986, officials made the announcement "The 32nd Annual Greater San Diego Science and Engineering Fair is officially closed." The responding cheer was one of joy and victory, for the participants, their teachers and their families.

Appendix A

ISEF Project Categories

BEHAVIORAL AND SOCIAL SCIENCES

Psychology, sociology, anthropology, archaeology, ethology, ethnology, linguistics, animal behavior (learned or instinctive), learning, perception, urban problems, reading problems, public opinion surveys, and educational testing, etc.

BIOCHEMISTRY

Molecular biology, molecular genetics, enzymes, photosynthesis, blood chemistry, protein chemistry, food chemistry, hormones, etc.

BOTANY

Agriculture, agronomy, horticulture, forestry, plant biorhythms, palynology, plant anatomy, plant taxonomy, plant physiology, plant pathology, plant genetics, hydroponics, algology, mycology, etc.

CHEMISTRY

Physical chemistry, organic chemistry (other than biochemistry), inorganic chemistry, materials, plastics, fuels, pesticides, metallurgy, soil chemistry, etc.

COMPUTER SCIENCE

New developments in hardware or software, information systems, computer systems organization, computer methodologies and data (including structures, encryption, coding, and information theory).

EARTH AND SPACE SCIENCES

Geology, geophysics, physical oceanography, meteorology, atmospheric physics, seismology, petroleum, geography, speleology, mineralogy, topography, optical astronomy, radio astronomy, astrophysics, etc.

ENGINEERING

Civil, mechanical, aeronautical, chemical, electrical, photographic, sound, automotive, marine, heating and refrigerating, transportation, environmental engineering, etc. Power transmission and generation, electronics, communications, architecture, bioengineering, lasers, etc.

ENVIRONMENTAL SCIENCES

Pollution (air, water, land), pollution sources and their control, waste disposal, impact studies, environmental alteration (heat, light, irrigation, erosion, etc.), ecology.

MATHEMATICS

Calculus, geometry, abstract algebra, number theory, statistics, complex analysis, probability, topology, logic, operations research, and other topics in pure and applied mathematics.

MEDICINE AND HEALTH

Medicine, dentistry, pharmacology, veterinary medicine, pathology, ophthalmology, nutrition, sanitation, pediatrics, dermatology, allergies, speech and hearing, optometry, etc.

MICROBIOLOGY

Bacteriology, virology, protozoology, fungal and bacterial genetics, yeast, etc.

PHYSICS

Solid state, optics, acoustics, particle, nuclear, atomic, plasma, superconductivity, fluid and gas dynamics, thermodynamics, semiconductors, magnetism, quantum mechanics, biophysics, etc.

ZOOLOGY

Animal genetics, ornithology, ichthyology, herpetology, entomology, animal ecology, anatomy, paleontology, cellular physiology, animal biorhythms, animal husbandry, cytology, histology, animal physiology, neurophysiology, invertebrate biology, etc.

Category Interpretations

Below are project areas about which questions frequently arise. It is included only to provide some basis for interpretation of the category descriptions.

INSTRUMENTS

The design and construction of a telescope, bubble chamber, laser, or other instrument would be Engineering if the design and construction were the primary purpose of the project. If a telescope were constructed, data gathered using the telescope, and an analysis presented, the project would be placed in Earth and Space Sciences.

MARINE BIOLOGY

Behavioral and Social Sciences (schooling of fish), Botany (marine algae), Zoology (sea urchins), or Environmental Sciences (plant and animal life of sea, river, pond).

FOSSILS

Botany (prehistoric plants), Chemistry (chemical composition of fossil shells), Earth and Space Sciences (geological ages), and Zoology (prehistoric animals).

ROCKETS

Chemistry (rocket fuels), Earth and Space Sciences (use of a rocket as a vehicle for meteorological instruments), Engineering (design of a rocket), or Physics (computing rocket trajectories). A project on the effects of rocket acceleration on mice would go in Medicine and Health.

GENETICS

Biochemistry (studies of DNA), Botany (hybridization), Microbiology (genetics of bacteria), or Zoology (fruit flies).

VITAMINS

Biochemistry (how the body deals with vitamins), Chemistry (analysis), and Medicine and Health (effects of vitamin deficiencies).

CRYSTALLOGRAPHY

Chemistry (crystal composition), Mathematics (symmetry), and Physics (lattice structure).

SPEECH AND HEARING

Behavioral and Social Sciences (reading problems), Engineering (hearing aids), Medicine and Health (speech defects), Physics (sound), Zoology (structure of the ear).

RADIOACTIVITY

Biochemistry, Botany, Medicine and Health, and Zoology could all involve the use of tracers. Earth and Space Sciences or Physics could involve the measurement of radioactivity. Engineering could involve design and construction of detection instruments.

SPACE RELATED PROJECTS

Note that many projects involving "space" do not go into Earth and Space Sciences. Botany (effects of zero G on plants), Medicine and Health (effects of G on human beings), Engineering (development of closed environmental system for space capsule).

COMPUTERS

If a computer is used as an instrument, the project should be considered for assignment to the area of basic science on which the project focuses. As examples: if the computer is used to calculate rocket trajectories, then it would be assigned to Physics, or if the computer is used to calculate estimates of heat generated from a specified inorganic chemical reaction, then it would be entered in Chemistry, or if the computer is used as a teaching aid, then it would be entered in Behavioral and Social Sciences.

Appendix B

Summary of Projects

BEHAVIORAL AND SOCIAL SCIENCES

Interaction Vs. Back-up: Testing Two Theories of Mouse Navigation by Ann Wood

This experiment was made to determine if mice, running through a T-maze to locate and reach food rewards, used a magnetic field as a backup to visual cues, or used both methods interactively. Although the test results strongly indicated the use of both methods interactively, the statistical level of significance, 0.05, was not reached in order to conclusively prove the hypothesis.

The Influence of Favorite Color, Sex, Age, Eye and Hair Color Upon Wardrobe Color, by Lan Do

In this project 419 Subjects, ages 5-76, completed surveys concerning their age, sex, ethnic origin, eye and hair color, favorite color, most popular color in wardrobe and eye and hair color of parents. Results indicated that color preference is sex related, with females preferring red, purple, and pink. The survey showed a preference for cool colors over warm colors as age increased. However, the survey results indicated that a person's favorite color had no relation to ethnic origin, eye color, or hair color. The most abundant wardrobe color is related more to the sex related preferences than hair or eye color, and is partially related to the stated favorite color.

Teenage Stress, by Kimberly Lea Quinlan

A study was done, using a questionnaire, in nine schools (four private and five public). There were a total of 378

responses. The analysis centered upon reasons for stress, which teens felt the most stress and how teens acted to relieve stress. Parents and tests were the major stressors. Public school students were less stressed than private school students, and females were more depressed or stressed by world problems than males. Alcohol and drugs were seen as responses to stress rather than their cause, but overall, private school students tended to eat while under stress and public school students used television and radio for relief. The results of the study correlated well with the ABC News survey of teenagers.

A Study of the Effects of Selected Carbohydrates on Alcohol Absorption and Behavioral Modification in Humans, by Catherine Jacobus

This experiment tested certain carbohydrates (fructose, glucose, sucrose, and saccharine) to determine if they had an effect on blood alcohol levels. Seventy subjects performed computer tests, sequence tests, dartboard tests and straight line tests. Blood samples were also drawn and tested for blood alcohol level, as well as glucose evaluation and SMAC profile. The experiment showed that glucose and fructose lowered blood alcohol and resulted in a better performance on the tests than saccharine. The sucrose group showed a higher blood alcohol level than the saccharine group. These results show a possibility that some carbohydrates serve as blocking agents that may modify alcohol absorption.

Mouse Appetite and Sound, by Michael Bruce

The object of this experiment was to determine whether sound affected the appetite of mice. The experiment was inspired by some classroom discussion dealing with subliminal advertising.

The subjects were divided into three experimental groups, one which was subjected to no unusual sound, one which was subjected to low frequency sound and one which was subjected to high frequency sound. The experimental groups were subjected to the sound for one hour each day during the experimental period.

Each day, an equal amount of food was given to each group. The next day, the amount remaining was measured before an entire measure was given.

At the end of the experimental period, the results showed that the control group ate less than either experimental group and that the group receiving low frequency sound ate less than that receiving the high frequency sound.

BIOCHEMISTRY

The Rabbit Test—Phases V & VI, by Elizabeth M. Robinson

This project was a continuation of earlier projects, which sought to control animal reproductive drives in the wild. Phase V of the project tested the ability of oral steroids to limit the reproductive capacity of adult male rabbits. In the experimental group, three of four males given steroids became sterile, while the fourth sired a healthy litter. Another experimental group was given human birth control pills. The results in this group showed that the human birth control pills produced a marked increase in male offspring.

A Comparison of Sugar Content in Soft Drinks by Jennifer L. Decker

This project tested our regular soft drinks and their diet equivalent, to determine the percentage of sugar in each. Coca-Cola, Diet Coke, 7-Up, Dr. Pepper, Diet Dr. Pepper, Pepsi, and Diet Pepsi were tested.

Hydrochloric acid was added to each sample to break the sugar down to reducing sugars. Benedict's solution was then added, and then standard glucose solutions were made in 5%, 10%, 15%, 20%, 25%, and 30%. Benedict's solution was added to these, as well.

Each solution was then heated, in order for the Benedict's solution to react. Each sample was then examined under the microscope, in order for Ms. Becker to count the black dots that represented the sugar. The number of dots in a 1/8 section were counted, both in the eight samples, as well as the standard percentages.

The results showed that regular soft drinks contain these approximate amounts of sugar: Dr. Pepper; 30%, Pepsi; 30%, Coke; 25% and 7-Up; 25%. No black dots were found in the diet drink samples.

The Synthesis and Biochemical Evaluation of Natural Estradiol Analogs in Cancer Chemotherapy by Latisha M. Keith

The objective of this process was to produce a synthetic estrogen analog. The application of this technology is important, because direct evidence has linked estrogen to the development of breast cancers. This study will continue with testing on live mice.

BOTANY

Radiation's Effect on the Plant Tissue Hormone, IAA by Jeffrey S. Morgheim

This experiment tested whether increased radiation (caused by a decrease in the ozone layer) altered the elongation properties of the plant regulating hormone IAA.

Oat seeds were germinated. An experimental group was placed in a solution containing IAA, which had been irradiated with laser radiation for 20 seconds. The control group was placed in IAA in its normal condition. Both groups were placed in the dark for 24 hours and measured for elongation. As the experiment progressed, the degree of IAA radiation in the experimental group was increased. The results did not indicate an increased elongation based on the increased radiation of IAA.

The Effect of Radiation on Bean Seeds and Seedlings, by John Stoffel

This project also tested the effects of radiation on seedlings. The experiment was carried out over several years. In the first phase, germinating and dormant seeds were exposed to varying degrees of radiation. The next phase exposed the buds to 6000 rads of radiation and in the final phase, the seedlings were exposed to 2000-6000 rads, at 7, 12, and 21 days of growth. Results showed that higher doses inhibit growth at a 2:1 ratio, but that wet seeds were twice as affected as dry seeds. The greatest effect appeared to occur at the period of greatest growth, 8-16 days after planting. The portion of the plant most affected was the terminal bud, followed by the roots and then the stem.

Mr. Stoffel concludes that this may be due to the rates of cell division at that time. He further concludes that due to the effects of radiation effects, in case of nuclear accident, knowledge of effects on plant life is essential to understanding the possible survival of the planet.

The Effect of Seed Size on Wheat—A Quantitative Study by Neil Ireland

This project attempted to measure the effect of seed size on the success of the crop. This is important, because wheat is a major food source throughout the world.

1000 seeds were measured with a micrometer and then categorized in 10 evenly spaced groups. Ten from each group were then grown in cotton wool in petri dishes and watered with a mineral solution. This was repeated for each of the experimental groups.

The number of seeds germinated, as well as the average

height of the groups of plants, were recorded. In addition, grazing was simulated by cutting off the plants at a specified height. Frost was simulated by freezing the seeds, and drought was simulated by water deprivation.

The results showed that a larger seed increased the initial growth rate and regrowth rate after grazing, as well as drought and frost resistance.

CHEMISTRY

Comparative Study of Raw and Pasteurized Milk from Different Dairies of Puerto Rico, by Francis Vazquez

This project was undertaken to determine the degree of milk contamination from different dairies, and the cause of such contamination. He tested both raw and pasteurized milk in quality control tests to determine acidity, lactometry and total solid percentage, as well as evaluating taste, odor, and appearance. The tests were also used to detect and measure the presence of bacteria and antibiotics.

The results showed that the consumers received milk with a 71% of quality. He determined that environmental qualities were responsible for 29% of the impurities. The presence of antibiotics, mainly penicillin, was probably due to penicillin used in the treatment of cows. Research was carried out using the facilities of the Puerto Rico Dairy Products Association.

Stop—You May Be Using The Wrong Detergent, by Gwen S. Pearce

This experiment tested eight detergents to determine the most effective in removing stains of grass, combined food, mud, and chocolate stains. All stains were cleaned in a water solution of each detergent and then evaluated and graphically represented. One detergent was rated best in terms of stain removal, yet conclusions were based on qualitative rather than quantitative analysis.

Nutra-Sweet in Diet Soda—A Study of Storage Temperature by Randall Shank

This project sought to study the degree of breakdown of Nutrasweet (Aspertame) in solution form. The experiment used diet soda samples, stored for a trial period of ten weeks, at 5, 15, 25, and 25 degrees C. Measurement was made using a High Performance Liquid Chromatograph.

The aspertame quantities decreased significantly at higher temperatures, with changes reflecting a logarithmic relationship. The data led to a formula that was used to produce data that closely followed the experimental data. The main "discovery" of the project was actually the formula, which may prove useful for determining the long term safety of aspertame.

COMPUTER SCIENCE

Computerized Speech Analysis & Recognition, by Carl T. Donath

This project was inspired by the need to input data into a computer without the need to type. The objective was to design an electronic circuit that could convert sounds into a series of numbers. First, the sounds, which are really changes in air pressure, are converted to electrical voltages via a microphone, and then turned into numbers by an analog-to-digital converter. Once the circuit was working, a program was written to interpret the numbers, searching for something that would distinguish one word from another. He finally drew an "envelope" around the graph that the word represents. These "envelopes" are stored in memory, and then used as the basis of comparison to subsequent "words" that are input.

Textlink by David Calabrese

This project developed the TEXTLINK system, which can "read" material from a typed or printed page and enter it into the computer memory. The purpose of this was to eliminate the need to re-key material for new word processors, data bases, etc., at a reasonable cost.

TEXTLINK works by using structural character recognition to identify characters within a computerized snapshot of a page. Digitizing equipment, which is relatively inexpensive, is used to take the photograph, "gives" the image to software, which recognizes the letter based on its unique structural components. Therefore, TEXTLINK achieved its objective by being an accessible, cost-effective text reader.

Computer System for The Handicapped, by D. Scott Smith

This project also attempted to use a computer system with "off the shelf" software geared for the handicapped. The object was to input data either with a speech synthesizer, or a joystick and/or touchpad. Three software programs were used, *English Based Interactive Language* (EBIL), a super high level language; *Words*, simple word processor; and *Talker*, a speech synthesizer.

Scott tested these programs over eight months, without using a keyboard or monitor, and found that the system could, in fact, be effectively used with alternate input and output devices.

Mendelevia by Angel de la Cruz

The purpose of this project was to create a computer program that would teach some of the basics about the elements. The basic assumption was that most of this material, such as names, symbols, atomic weights, etc., was rather boring.

Therefore, the objective was, rather than creating educational software, to write a "game" program, which students would enjoy, and therefore continue playing and learning chemistry in the process.

EARTH & SPACE SCIENCES

The Shaking Earth, The Burning Sky by Mike Iritz

This project attempted to correlate the incidence of earthquakes of magnitude 6.0 and above, and 7.5 and above, with incidence of high solar activity. Data for the years 1937 through 1975 was collected, graphed, and correlated using the Coefficient Correlation. Although graphically, a correlation was shown, statistically, the hypothesis was disproved.

The Effects of Microgravity on Embryogenesis by John Chin-Hung Leu

This project studied the difference in the development and spawning of amphibian eggs under simulated hypogravity conditions. The experiment showed that in the control group, more than 60% of the eggs were shown to develop normally with respect to size, form, etc. However, in the experimental group, development of the eggs was delayed, and also structure of the eggs was abnormal. Mr. Leu plans further studies to analyze the precise factors that lead to the developmental delay, and examine the synthesis of chemicals within the eggs.

Variation in Crystal Habit of Pyrite from Otoge Mine, by Koji Morimoto.

This project studied the different patterns of growth of crystal formations on pyrite taken from different portions of the same mine. About 20,000 samples of pyritic crystal were collected and classified. After analysis and study of the samples relative to their source, the conclusion reached was that the location of its formation, particularly as rates of cooling varied in different parts of the mine, was a fundamental cause of the variances in rate and type of crystallization.

Hohokam Pottery Porosity, by Andrew J. Wolf

This project involved testing the porosity of pottery shards, to determine if this study would indicate that pottery was produced differently depending on its intended use.

The hypothesis was that the pots would be more porous because they were used as water containers, and a more porous container would permit a film of water to form on the outside, creating an evaporating effect. The test was done by weighing the shards dry, then submerging them in boiling water and weighing them again.

The testing proved the hypothesis incorrect. Mr. Wolf speculates that this may be because bowls may have been used for cooking, as well as for storage, and that in cooking, more porous materials would heat more evenly.

ENGINEERING

Design, Construction & Operation of a Continuous/Pulsed Ion Beam Accelerator, by Stephen M. Jacobs

The objective of this project was to build a small accelerator for use in high energy particle physics. The idea was derived from previous projects, which dealt with various aspects of the atom, namely the electron cloud and the nucleus.

Materials for the project, such as small parts and electrical supplies were purchased at shops, although a great many more specialized pieces of equipment were borrowed. Household materials were also used. The purpose was to develop the best design, using "common" materials, for safe and efficient use.

The main problems encountered were the proper focusing of the ion beam and the minimizing of the gassing of the device.

Fly Ash Foam—A Method to Create A Ceramic Material from The Waste Products of Coal Incineration Through The Use of Microwaves by Caroline K. Horton

The objective of this project was to create a useful product, a strong lightweight foam, from the fly ash, which was a waste product from coal-fired heating and electrical generating facilities.

To produce the foam, Horton developed a solution to add to the fly ash and then microwaved it, producing a strong ceramic foam. To shape the foam, she produced fiberglass molds in which to encase the foam solution prior to the microwave process. Horton began her research at the Los Alamos National Laboratories, which produced its own fly ash foam. However, the foam produced by her method was stronger and more durable than that developed at Los Alamos. Therefore, because of its commercial application in construction and insulation, she is currently applying for a patent for this new method of producing the foam.

Coil Beam Magnetic Accelerator, by Michael James

The objective of this project was the development of a coil gun, which fires at great speed with low input power, by use of induced magnetic fields. This project was triggered both by previous research as well as an interest in SDI research.

It was determined that to yield high input velocity and low input power, large capacitors would need to be used to supply the magnetic coils with current. Also, the wire gauge, length of coil and number of wraps were optimized to supply highest acceleration. Optical triggers were used to fire each stage, making the systems self-correcting. Five coil stages yielded a projective velocity of 28.68 meters per second.

ENVIRONMENTAL SCIENCES

Rehabilitation of Ground Water Aquifers, by Terri Newport

This project grew from a concern with hazardous waste sites, particularly how they affect ground water. The objective of the project was to develop and demonstrate methods of identifying, isolating, and rehabilitating ground water aquifers.

Newport accomplished this using five physical models that would duplicate actual conditions. These models were constructed of plexiglas, to permit easy observation, and fitted with injection and pumping wells, which allowed for sampling the contamination concentration. In each instance, a rehabilitation strategy was developed and demonstrated, identifying its effectiveness and limitations.

This project grew out of work on a prior year's project, *Ground Water Pollution—Identification, Isolation and Rehabilitation.*

Acid Rain—Causes, Effects, and Cures, A Seven Year Study, by Leslie S. Thomas

This seven year project attempted to find the causes and effects of acid rain, as well as a method of correcting the damage inflicted by acid rain. The final phase dealt with a method of implementing corrective measures previously discovered. The neutralizing agent was tested in two acidified ponds in the Adirondaks. Once the ponds were neutralized, they were restocked with native fish and plant life. After eight generations, these species are still thriving.

Controversial Issue—Are EPA Toxic Waste Standards Safe? by Eric J. Felt

This project attempted to prove whether the EPA's safe levels of toxic waste were, indeed, safe. To conduct the experiment, Mr. Felt used the acceptable levels of three different types of toxic waste, herbicides, solvents, and heavy metals. He then tested planaria for their ability to regenerate as well as perform conditioning tests.

Although each of the wastes produced different impairment levels, using the Chi-square method of validation, it was shown that the "acceptable" levels of each waste, did, in fact, affect the regenerative and conditioning abilities of planaria.

Can A Charcoal Filter Reduce The Effects of Automobile Exhausts on Chick Embryos? by Libby Haines

This project was based on prior research, which proved the harmful effects of fumes, from both leaded and unleaded gasolines, upon chick embryos. In this project, a filter was developed in a attempt to lessen the effects of the fumes, using plastic pile with charcoal. These filters were placed on cars with leaded and unleaded gasoline, and then introduced into zip-lock bags containing chick embryos. Unfiltered fumes, as well as a control group subjected to no fumes, were also part of the test.

The results showed increased growth of the embryos exposed to the filtered fumes, as opposed to those exposed to non-filtered fumes.

An easy, inexpensive method of reducing pollution is the possible development of a "throw-away" charcoal filter to be attached to the end of a gasoline exhaust pipe.

MATHEMATICS

Blackjack by William Chen

Chen's project attempted to expand on or improve the various blackjack systems available, which were based on counting cards. He determined that the existing systems yielded a return ranging from 1.1% to 1.3%. He altered Julian Braun's system to count aces, but was unable to appreciably improve the rate of return. Mr. Chen concluded that the only effective systems were based on counting cards, and even when those systems are adhered to, gain is at a slow rate. The tests were run using Turbo Pascal.

An Original Method for Solving Tangential Circle Systems by Jennifer Wood

This project attempted to find the solution for a fourth tangent circle in a system when the other three were given. Although there have been solutions using quadratic equations, they were very lengthy processes. Ms. Wood's solution is shorter and more efficient. She developed a numerator, which is the product of the first two radii divided by the third, then normalizing the results. Subsequently, all additional circles could be found by sequentially adding all denominators to the sequence.

Wood's results were reviewed and verified by a professor of mathematics and engineering at the University of Southern California.

It's Not Cartesian—A Study of A New Graphing System, by Christopher C. Finger

Finger developed the Triangle graphing system in which each pair of coordinates defines a segment that joins the axes. This is unlike the Cartesian system where each pair of coordinates defines a point. The research for this project included studying the slopes of segments, conic sections, and inverse equations, as well as solving simultaneous equations to determine when segments cross.

Proceeding with the study, Mr. Finger treated the slope

of a segment as though it were a Cartesian line and proceeded to study the slopes in various equations. Study and research pointed out two flaws in the system. First, any segment with one coordinate of zero cannot be seen because it lies on an axis. Secondly, a step between x values must be specified, because with an infinite step, single segments cannot be seen.

MEDICINE AND HEALTH

Malaria Vectors—Eight New Species Implicated by Elke Baker

This project grew out of Baker's interest in tropical diseases and the opportunity to work in a tropical medicine laboratory over the summer. She was trained in the laboratory techniques while there and worked closely with a supervisor on many projects.

The objective of the project was the identification of the species of mosquitos that are infected with the human malaria species and the transmission in a particular geographic area. By testing the ELISA (enzyme linked immunosorbent assays), which detects sporozoites in mosquitos, on samples of 13 species of mosquitos, Baker uncovered the possibility that 8 new species of mosquito were implicated in the disease's transmission.

One advantage was Baker's research at the Walter Reed Army Institute of Research. A disadvantage, however, was that her specimens were two year old freeze-dried mosquitos, without accurate records of how, when, and where they were collected.

Immune Evaluation of HHT by Lisa J. Chin

HHT is a cephalotazine ester from several widely spread evergreen trees in China. Since HHT is under investigation for treatment in some types of leukemias, this project attempted to determine whether HHT was an immunosuppresive antitumor agent. Concentrations of HHT were added to a culture that contained human lymphocytes. All concentrations of HHT tested were found to be inhibitory to lymphocyte proliferation.

Stuck on You—A Study of Dental Adhesives by Stuart Allen

This project attempted to prove that substances that are high in sugar, such as cola or Kool-Aid, weaken the bond between teeth and braces, or other orthodontic appliances. To test this hypothesis, four groups of false teeth with orthodontic bands cemented on were soaked in four substances, cola, Kool-Aid, beer, and water. They were then subjected to 2250 grams of pull using a spring scale. Only one of the bonds came apart, on the first day of the experiment. Therefore, the hypothesis could not be proven by the experiment.

Computer Simulation of The Plasma Membrane's Transport Functions, by Dennis Petrocelli

This experiment attempted to create a computer simulation of the transport function of the plasma membranes. Two BASIC programs were created to accomplish this. The Develop program designed membranes by adding functional units to a lipid bilayer and the Membrane program simulated diffusion and facilitated diffusion across the membranes created with Develop. To validate the process, four experiments were done that investigated different aspects of transport. After the experiments, the simulations proved to be accurate, and therefore of potential use in the analysis needed to cure human diseases of the membrane.

The Canine Allergy-Obesity Connection by Missi J. Wilkenfeld

This project attempted to determine whether allergies predisposed individuals to obesity. The hypothesis was tested using a sample of 50 dogs. A set of criteria was developed to indicate an allergic condition, and the dogs were classified as allergic or non-allergic based on the results of the questionnaire. Six of the dogs were classified as normal and 44 were considered allergic.

Each dog's body weight was then compared to established norms. A pinch test was also done on each dog to determine the body fat content. If the dog's weight was 15% above average, it was classified as overweight, and if it was 20% above average, it was classified as obese.

Testing showed that the majority of dogs classified as allergic had weight problems. Of the 44 allergic dogs, 30 (68%) were either overweight or obese. Of the six normal dogs, only 2, representing 33%, were overweight or obese. Wilkenfeld hypothesized that the allergies slowed the metabolism. Hence, the results of this experiment indicated that allergy treatment might alleviate weight problems.

MICROBIOLOGY

The Curing Effect of Ultraviolet Light on The F'Proline+ Lactose+ Strain of Escherichia Coli, by Jessica Wertlieb

Wertleib was introduced to this topic by a professor of genetics at San Diego State University. Her main sources of information were the professor himself and several local university libraries. This project was of particular interest because although it was successful at winning several awards, Wertlieb overhauled her procedures at least seven times in the course of the experiment, because there were problems with faulty equipment, as well as difficulties in getting the organisms to grow properly.

The Differentiation between Human and Cow E. Coli, by Robin Rowe

Escherichia Coli is used to indicate fecal pollution when

fresh water is tested. However, when this substance is found, there has been no practical method of distinguishing between human and cow E.coli, therefore making it difficult to determine the cause of the pollution, and hence take steps to curtail it. This research consisted of two phases, one which tested the E.coli in five different sugars, and a second phase in which the sugars, plus brom cresol purple broth was used. The procedure used in phase two enabled absolute differentiation between human and cow E.coli.

Effect of pH Upon The Culture—Penicillin Notatum, by Shawn Bray

Studies show that penicillin notatum colonizes best at a pH of 7.0, and that growth is difficult above pH 8.0 or below pH 5.3. Because the penicillin mold grows on bread, yet can only colonize within the stated ranges, this project concludes that mold can be prevented by adjusting the pH level of bread.

Jalapeno Peppers, Garlic and Sage as Natural Preservatives and Germicidal Agents by Tracy M. Holje

This project tested whether a home remedy, consisting of jalapeno peppers, garlic and sage, work as a preservative on meat, and also can be used as a germicidal agent on cultures of mold and bacteria.

Both chemical and home remedies were applied and then bacteria was added to nutrient agar on the plate. After 24 hours, the samples were examined and the bacterial colony count recorded. The tests were repeated several times, each time, isolating the best preservatives and germicidal agents for further testing.

Finally, benzoic acid was proven to be the best chemical preservative, and whole garlic the best natural preservative.

PHYSICS

Obtaining Kinetic Energy through Heat, by Billy Scott

The objective of this project was to prove the theory that the temperature of a body is equivalent to the average kinetic energy of its constituent particles of molecules. The hypothesis was that a change in the temperature of the body would result in a change in the kinetic energy. Scott experimented with stretched opposing elastic bands between two points with equal and opposite force. When one side of the band was heated the entire band contracted, stretching the corresponding band an equivalent amount, hence resulting in increased energy. As long as the bands were kept in motion, the potential energy created was changed to kinetic energy.

To conduct the experiment, a device consisting of a bicycle wheel with existing hardware removed was used and elastic bands were symmetrically placed throughout the circumference of the wheel. A light bulb was used as the heat source.

What Happens When The Length of A Pendulum Is Increased, by Nicola Kean

This project examined the gravitational differences resulting in the different lengths of a pendulum. This project was interesting because Kean looked for a simple project with easily interpretable results, yet needed to delve into research in an unfamiliar field. However, she felt that she learned a great deal about physics and the laws of gravity as a result of the project.

In addition, Kean used a computer for her display, including graphs, which generated attention at the various fairs in which she exhibited.

Visual Reversal of Photographic Negatives, by Maryanne Large

The basis of this project was the analysis of obscure optical effects when black and white photographic film was viewed from the same side as the light source. These effects were the scattering effect, resulting in a positive image and a reflectance effect, producing a negative image. Testing was done to quantitatively determine the effects of density development and the angle of incident light on the images. Qualitative analysis was carried out using colored filters and electron micrographs were also made to aid the understanding of the importance of the structure of the film itself to the reflectance negative.

The project was inspired when Large accidentally noticed the optical effects when developing and printing photographs at home.

Real-Time Transmission of Holographic Interferometry-Determination of Minute Refraction Index Changes for Fluid Aerodynamic Applications, by Mark David Owens

The objective of this research was to design, construct, modify, and apply a real-time transmission holographic interferometry system, capable of measuring minute changes in refraction indexes to facilitate the visualization of water flow past a test specimen, thus showing the water's relationship to the object.

Most of the device was built using scrap materials, for example, an old water pump from a washing machine. Over forty devices were built and tested, until one was found to satisfy the project objective. This design may act as an alternative to standard water channels and wind tunnels for military, medical, chemical, and engineering applications.

What Is The Best Concave Design for Water Skis, by Ronald E. Thompson

First, Thompson determined that the concave design

Science Fair: Developing a Successful and Fun Project

for water skis was superior to flat design, but also discovered that there hadn't been any design innovations. He then created over 30 new concave ski designs, then tested them to narrow the selection to two designs.

These two were then tested against the concave ski design manufactured presently, all 1/5 scale size. Tests included surface tension, friction, and stability, conducted at a speed of 22.5 mph.

One of the designs proved superior to those currently in use, enabling easier learning and proficiency with the sport. A secondary product of the research applied the design to boat hulls, creating one combining the strength of a deep V hull with the speed of a tunnel hull.

ZOOLOGY

The Role of Amoebocytes in Zinc Uptake and Transport in The Snail Helix Aspera, by Ben Cheng

The objective of this experiment was to determine the role of mobile cells termed amoebocytes, which are amoeba-like cells, in the uptake and transport of metal particles in molluscs. These cells can accumulate these metal particles and transport them to other organs, such as the kidney.

To determine the role of amoebocytes in this process, they were isolated from the blood of the land snail, Helix Aspera, and then incubated with 65-zinc, because it plays an essential role in biological systems, and 109-cadmium, which does not participate in any known biological processes. The partitioning of the metal between plasma and amoebocytes was monitored in order to determine the uptake of soluble metals by ameobocytes.

The results showed that the amoebocytes apparently do not play an important role in the uptake and transportation of zinc and cadmium in the snail, Helix aspera.

Carbon Dioxide Is Necessary for Coral Growth, by Daniel S. Cheng

This experiment was conducted with two aquariums, one containing fish and one without fish, each containing coral growth. Observations and measurement on the amount of growth of the coral was made over a period of 14 days. After analysis of the data, the conclusion was that carbon dioxide, in the experimental case, generated by the fish, was necessary to the growth and survival of the coral. This was a rather expensive project, requiring the purchase of two aquaria plus four pumps. Mr. Cheng also encountered some difficulty collecting identical species and sizes of coral for the experiment.

A Study of The Wasp—The Mechanism of Determination of The Sexual Function in Vespula Lawisii, by Soya Akira

This project attempted to determine why queen wasps can oviposit, or lay eggs, while workers, which are also female, cannot. There are some basic differences in the reproductive organs, as follows. First, the ovary of the queen develops well, while those of the workers do not. Second, the ollicular egg cells of the workers do not grow as powerfully as that of the queen. Finally, oocytes with yolk granules can be observed in the queen, but not in the workers. These differences were determined by the differences in the amount of hormonal secretion in the queen and the workers. This was the most important factor in determining the sexual function of the queen wasp and the workers.

Drugs Weave Patterns, by Doug S. Fay

This experiment showed the effects of caffeine, alcohol, and nicotine on the geometric pattern of a spider's web. Because these spiders build webs daily and consume them at night, dilute solutions of the chemicals were misted on normal webs, and photographs were taken the next day when new webs had been woven. Drug influenced webs differed drastically from normal webs, indicating that the drugs affected the spider's ability to weave its web, and hence, its ability to function normally.

Appendix C

Glenn T. Seaborg Nobel Prize Visit Awards Winners' Abstracts

37TH INTERNATIONAL SCIENCE AND ENGINEERING FAIR, FORT WORTH, TX., 1986

A New Algorithm for Optimal Route Selection, **Pamela S. Heady**

In the future, robots will become more mobile. However, to be really useful, they must be able to execute optimizing commands. The Travelling Salesman problem involves such an optimizing command. It requires finding the shortest simple path connecting a given set of destinations. An N-station Travelling Salesman problem has $(N-1)!$ possible paths, and for N greater than about 20, even the fastest computer cannot do an exhaustive search in a reasonable time. By using the most advanced search algorithms, problems with about 30 destinations can be solved, but still only when run for weeks on a Cray computer.

My goal was to find a very fast, very accurate approximate solution to the Travelling Salesman problem. I hypothesized that useful results could be obtained by eliminating the requirement of reaching an exact solution. This is a practical goal because in many situations a path that is only slightly longer than the best will be considered adequate.

By using geometric analysis I found four principles which allow 30 point problems to be solved in about 7 seconds. The 48 state Travelling Salesman problem was solved in about 14 seconds. For N less than 150, my program runs in polynomial time (exponent = 1). Where I could check the solution, it was always exact, but I can show that certain cases will not be precisely correct.

I call this a "robotics solution," because it is much more accurate and far faster than what a human can do, and can be programmed into a small, mobile computer.

36TH INTERNATIONAL SCIENCE AND ENGINEERING FAIR, SHREVEPORT—BOSSIER CITY LA., 1985

A Three-Year Study of Computerized Differential Identification Using the Dental Configurations and Bitemarks of Homo Sapiens, **Sheryl L. Ames**

Because fingerprints cannot always be used to identify individuals, modern science and technology is now looking at the teeth to provide a new means of differential identification that will be a more permanent record of an individual's identity. The purpose of this research is to develop an inexpensive and reproducible computerized system that utilizes the dental bitemarks of an individual for differential identification. A bite wafer was developed to register an individual's bitemarks and the bitemarks of 183 test subjects from elementary schools were taken. A computer program was then written to convert the bitemarks into numerical values that are used to identify unknown or missing persons. A comparison of the differences in the bitemarks of the test subjects after one year demonstrated the uniqueness and validity of these values as differentiating characteristics. Because of the uniqueness and relative accuracy of this system, several identification agencies and professionals in the field of forensic dentistry have expressed an interest in the system.

The Development of a System for Acquisition of 3-Dimensional Images, **Eugene Sargent**

Sight is our most relied on sense. We use it in almost

everything we do. In order for a device possessing artificial intelligence to be able to interact with its environment as we do, it would need some way of seeing the things around it. It would need a system for getting three-dimensional images of its surroundings.

In my project, I developed a system, to be used by a computer, for getting digitized images of three-dimensional objects. My system utilized a small neon laser and a system of motors and mirrors to scan the surface of the objects. After a great deal of experimentation with photodetectors, I developed a successful array of photodetectors to gauge the distance of the laser spot. It utilized three specially modified cadmium sulfide cells and a 216 fiberoptic light guide to collect and distribute the scattered laser light. I designed and built a circuit to process the outputs of the Cds array, which I interfaced to a Commodore 64 by adding three input ports to the expansion port of the computer. I designed a program to decode the output of the array and to store it on disks. I developed a system of assembly language graphics routines, which plotted the image in simulated 3-D and removed the hidden lines from the plotted image.

Although my system is still a prototype, it has produced surprisingly good results. It has a fairly low resolution and a limited range, but the data is consistent and free of noise, I plan to improve the recognition and accuracy of my system and to go on to image recognition.

35TH INTERNATIONAL SCIENCE AND ENGINEERING FAIR, COLUMBUS, OH 1984

Net Transfers of Lipids Between Plasma Lipoproteins: A New Biochemical Index for Detecting Human Coronary Artery Disease, Ann R. Davis

Purpose. To devise a new procedure for measuring simultaneous in vitro net transfer pathways of lipids between lipoproteins of human fresh plasma and to determine whether any differences in these pathways occur in middle-aged males with coronary artery disease compared to age-matched normal healthy male humans.

Procedure. Following informed consent, 10 ml blood samples were obtained from 12 middle-aged male patients with coronary artery disease and 12 age-matched normal male human volunteers. A new procedure was developed for measuring the extent of net transfer of unsterified cholesterol, esterified cholesterol, and triglycerides between human plasma lipoproteins during a four hour incubation of fresh plasma at 37 degrees C. This new procedure utilized cellulose acetate electrophoretic separation of plasma lipoproteins coupled with presently developed new "sandwich" methods for enzymatic staining of plasma lipoprotein-lipid bands.

Results. A new biochemical index number for coronary artery disease, named the plasma lipoprotein-lipid transfer (PLLT) index, has been calculated from six plasma lipoprotein-lipid net transfer pathways, which were found to be statistically different in middle-aged male patients with coronary artery disease compared to age-matched normal healthy male volunteers. The PLLT index averaged 9.27 + 2.38 (or 9.27 – 2.38 mean + or – S.E.M.) in 12 male humans with obstructive coronary artery disease and 0.54 + 0.25 (or 0.54 – 0.25) in 12 age-matched normal male humans, corresponding to a 1,617% increase in coronary artery disease (p < 0.005), with a 92% correct classification for the diagnosis of coronary artery disease.

Conclusions. The present data suggest that the PLLT index may have a possible use as an initial screening marker for coronary artery disease.

Immunity to Herpes Simplex Virus I Conferred by Genetically Restricted Variants and Immunoactivating Factors, Timothy A. Thrailkill

Active immunity to herpes simplex virus 1 has been conferred with the use of an experimentally effective live vaccine. Vaccine samples were composed of inductor-exposed variants of parental H166 + in aerosolized form for oral inhalation. Variant populations were assured of monostrain purity by monoclonal antibody selections. Variants were screened for in vitro avirrulence, in vivo apathogenicity, antigenicity and immunogenicity. A prototypical vaccine affirmed sufficient attenuation and a potentiated vaccine confirmed immunogenic potential. Studies of nodal infectivity indicate primary replication in the oral cavity, with associated neural invasion, and secondary replication in splenic and helatic tissues. Proposed genotypes based on phenotypic data gave insight as to mechanisms of the abortive viral infection.

Final vaccine experiments explored immunizing regimens, Immunogenicity indices were designed on nomographs. Protocols on murine subjects resulted in full protection versus latent recurrence. Protection versus latency was by 1:1 cellular colonization in the trigeminal ganglia. Immunity was polytypic in nature in respect to other HSV-1 strains. Potentiated Cai 166 samples were safe in immunocompromised mice. Rabbit testing demonstrated the effectiveness of the vaccine. Overall results support the hypothesis that this vaccine is ready for human subject trials.

Recent studies indicate nontransmissibility of variants from vaccinated to unvaccinated individuals, probably due to reduced chance of morbid recrudescence by variants. Communal efficacy of the PL166' vaccine was proven resistant to acquiring contagious infection. Future research involving occult latency is in progress.

34TH INTERNATIONAL SCIENCE AND ENGINEERING FAIR, ALBUQUERQUE, NM, 1983

A Continuing Analysis of Alligator and Crocodile Tooth Microstructures and Their Potential for Evaluating Higher Systematics, Laura Trauth

Although naturally shed alligator and crocodile teeth are

some of the most commonly found Cretaceous fossils of North America, they have little taxonomic importance unless they are still lodged within a much rarer fossil skull and identified through skull fenestration. Results of a previous phase of this project suggested that internal ring structures of such teeth might be genetic invariants and, thus, have taxonomic value.

This phase of the project devises a means by which individual teeth may be quantitatively identified through their dentine ring systems, making the presence of a skull unnecessary. The teeth used in this study were known modern teeth from two families of crocodilians and unidentified fossil teeth. The teeth were evaluated mathematically, with the equations describing the curves of the dentine rings as the discriminating factors. The equations were found to differ at the species level for modern teeth. The fossil teeth were less easily classified because of their unknown character and interfering subring structures.

These structures were found to have possible counterparts in modern teeth soaked in a protease solution. The fossil teeth and the protease treated modern teeth were less fracture resistant than non-treated modern teeth—suggesting that protein is a strength factor. This hypothesis requires further research.

Thus, as well as providing a model for identification of individual crocodilian teeth, this study suggests that some fundamental aspects of tooth structure may be better understood through study of tooth proteins.

Tip Vortex Propulsion/A New Approach, Jonathan Santos

The tip vortex is a tornado-like disturbance, which is formed by the lower pressure lift zone drawing in the higher pressure air around the wing tip. I was curious to research possible ways to utilize the vortex instead of wastefully trying to eliminate it.

Researching this, I read an article on the unusual outstretched wing-tip feathers on larger soaring birds, and how those feathers detect the vortex for propulsion. I started my project with the idea of simply designing similar fins to be applied to the aircraft for the reduction in fuel consumption.

After three months of research in a home-made wind tunnel, I determined that this type of system would be inefficient and impractical due to the fact that the sails (except the first) lie in the turbulent wake of the sail before it. I then devised a way to incorporate the several separate fins into a single lifting surface. The technique to place this single sail at its most efficient angle of attack around the three-dimensional vortex was experimentally derived.

My concluding experiments determined that the "mono-sail" produces increased flight efficiency of 27 percent.

Glossary

abstract—A short summary of the main points of a project. This is normally between 200-250 words in length.

analyzed data—Data derived from raw and smooth data, from which conclusions can be drawn.

conclusion—Interpretation based on outcome of results and answering the question or comparison suggested by purpose.

control group—Identical to the experimental group in all aspects, except that no variables are applied. This represents the test group that has all variables standardized and forms the basis for comparison.

controls—This represents factors that are not to be changed or variables that are to be controlled. **Do not** confuse with control group.

dependent variable—The factor that changes as a result of altering the independent variable. Also, the change in events or results linked and controlled by another factor that has also been changed.

experiment—A planned investigation to determine the outcome that would arise from changing a variable or from changing "natural" conditions.

experimental group—A group of subjects to which independent or experimental variables are applied.

experimental variable—See *independent variable.*

graphs—Illustrated form of presenting raw, smooth, or analyzed data.

hypothesis—Statement of an idea that can be tested experimentally, based upon research. States what experimenter believes will happen as a result of the experiment.

independent variable—The item, quantity, or condition that is altered to observe what will happen; something that can be changed in an experiment without causing a change in other variables.

interpretation—One's personal viewpoint based on the data. This can be based on either qualitative or quantitative analysis, and may become a part of the project's conclusion.

materials—All items used in the course of the experiment.

measured variable—See *dependent variable.*

observation—What one sees in the course of the experiment. Observations are often incorporated into raw data.

procedures—Steps that must be followed to perform an experiment.

proposal—The planned procedure for experimentation.

qualitative analysis—Analysis made subjectively, without measurement.

quantitative analysis—Analysis made objectively, with measurement devices.

question (or problem)—That which forms the basis of the hypothesis and hence is the objective of the experiment.

raw data—Logs, tables, and graphs that represent data as it is collected in the course of the experiment.

research—The process of learning facts or prior theories on a subject by reviewing existing sources of information.

results—Graphs and tables that represent raw, smooth, and analyzed data.

scientific method—Manner of conducting an experiment, using valid subjects, variables and controls, and accurately recording results.

smooth data—Tables or graphs where all the averages, totals, or percentages are placed. These may combine the information from raw data tables and graphs.

tables—Written form of presenting raw, smooth, or analyzed data.

variable—A condition that is changed to test the hypothesis or a condition that changes as a result of testing the hypothesis.

Bibliography

Horowitz, Lois, *Knowing Where to Look*. Cincinnati, OH: Writers Digest Books, 1984

Science Service, *Abstracts—37th International Science and Engineering Fair*. Washington, D.C., 1986

Science Service, *Abstracts—36th International Science and Engineering Fair*. Washington, D.C., 1985

Science Service, *Abstracts—35th International Science and Engineering Fair*. Washington, D.C., 1984

Science Service, *Abstracts—34th International Science and Engineering Fair*. Washington, D.C., 1983

Seeber, Edward D., *A Style Manual for Students—Based on the MLA Style Sheet*. Bloomington, IN: Indiana University Press, 1964

Strunk, William and White, E. B., *The Elements of Style*. New York, NY: Macmillan Publishing Company, 1979

Stoltzfus, John C., and Young, Dr. Morris N., *The Complete Guide To Science Fair Competition*. New York, NY: Hawthorn Books, Inc., 1972

Van Deman, Barry A. and McDonald, Ed *Nuts & Bolts, A Matter of Fact Guide to Science Fair Projects*. Harwood Heights, IL: The Science Man Press, 1980

Index

Other Bestsellers From TAB

"A wonderful gift for teens, this workbook brings the wisdom and acceptance of a wise and loving grandparent together with the feeling of having a best friend who really understands. Building upon the *Mindful Self-Compassion* and *Making Friends with Yourself* curricula, Karen provides teens with a path toward navigating the challenges of adolescence and developing an inner resource of wisdom and compassion. This workbook can change the course of teenagers' lives by providing the emotional resilience to get through challenges and pursue their dreams. Teens need never feel alone again."

> —**Michelle Becker, MA**, licensed marriage and family therapist, compassion teacher, cofounder of MSC Teacher Training, and founder of the *Compassion for Couples* program

"Wow! This book gets right to the heart of self-compassion, offering life-changing exercises in the easiest possible way. Written by the top expert on teens and self-compassion, it is based on solid research and the experience of thousands of people whose lives were transformed by the practices. I'll be recommending this book not only to teens, but also to the teenager in each of us."

> —**Christopher Germer, PhD**, lecturer in psychiatry at Harvard Medical School and author of *The Mindful Path to Self-Compassion*

"Teens, this workbook is a fun way of exploring how to more deeply know and care for yourself, your friends, and family. The authors offer lots of creative ways to explore your inner life, get to know yourself better, and take control of your life to build a caring and compassionate world."

> —**Mark Greenberg, PhD**, Bennett Endowed Chair in Prevention Research at Penn State, and author of over 350 journal articles and book chapters on prevention for mental health concerns and the promotion of well-being

"By learning the art of befriending yourself, you can become at peace in the world—with yourself, your friends, family, and peers. It sounds simple, but teens know it's not always so easy these days. But I know that you can do it, and this wonderful book shows you how."

> —**Dzung X. Vo, MD**, author of *The Mindful Teen*

"As if everyday life isn't challenging enough, most teens add to their struggles by unnecessarily judging themselves when problems and challenges arise. By learning to treat themselves with the same kindness and compassion that they show to their friends and loved ones, teenagers can build confidence, reduce their stress and unhappiness, and face life's challenges with greater equanimity. Bluth's exceptionally engaging and accessible book should be required reading for all teens (and their parents)."

—**Mark Leary, PhD**, professor of psychology and neuroscience at Duke University, and author of *The Curse of the Self*

"*The Self-Compassion Workbook for Teens* is highly engaging, realistic, and wise. Bluth has anchored the applications of self-compassion to the common and highly stressful experiences of adolescents, as they negotiate family, peer, and school pressures. There is a wonderful balance of mindfulness, self-kindness, and common humanity that should enable young people both to manage stress and to develop greater empathy for others. The distinction between self-esteem and self-compassion may be the most important contribution of the workbook to preventing depression in adolescents."

—**John F. Curry, PhD, ABPP**, professor in the department of psychiatry and behavioral sciences, and department of psychology and neuroscience at Duke University

"This book offers powerful skills for facing the daily challenges of life as a modern teenager. The beauty of these skills is that they help us when we are feeling most alone, useless, and hopeless. It is a huge relief to know that there are some very simple, easy things that we can do to support ourselves when we are feeling crappy and unhappy. So, if you are a typical teenager and feel like this sometimes or often, open this book, and begin learning how to be compassionate with yourself."

—**Amy Saltzman MD**, author of *A Still Quiet Place for Teens*

"In this engaging workbook, Karen Bluth provides teenagers with a valuable road map to their minds and hearts as they navigate the ups and downs of adolescence. In her warm, authentic, personal voice, she draws teens in using art, music, writing, photography, humor, and creative activities to help them connect to this wisdom on a deeply personal level. She grounds this book in the science of compassion and mindfulness, and translates it into practices that resonate with teenagers' lived experiences. This workbook is a terrific resource for everyone—but especially for young people struggling with the challenges of self-criticism and anxiety. Bluth not only teaches 'about' compassion; she communicates compassion through her openhearted message to teens everywhere: *you are not alone.*"

> —**Trish Broderick, PhD**, clinical psychologist and research associate at the Edna Bennett Pierce Prevention Research Center at The Pennsylvania State University, author of *Learning to Breathe*, and coauthor of *The Life Span*

"The teen years can sometimes seem like an indecipherable and sometimes frightening mystery, for parents AND for teens. Karen Bluth has brought her wealth of knowledge and wisdom to untangling the mystery and unlocking the truly transformative power of practicing self-compassion. This is a delightful and, above all, practical workbook for teens to discover their capacity to not only be aware (mindful) of themselves and their thoughts and feelings, but also to locate their innate ability to be kind to themselves when they face the inevitable challenges, bumps in the road, and feelings of inadequacy that are a common part of becoming an adult. Teens and parents alike will love this book and get so much out of the practices it teaches."

> —**Steven D. Hickman, PsyD**, associate clinical professor at the University of California San Diego School of Medicine; executive director for the Center for Mindful Self-Compassion; and founding director of the UCSD Center for Mindfulness

the self-compassion workbook for teens

mindfulness & compassion skills
to overcome self-criticism
& embrace who you are

KAREN BLUTH, PhD

Instant Help Books
An Imprint of New Harbinger Publications, Inc.

Distributed in Canada by Raincoast Books

Copyright © 2017 by Karen Bluth

        Instant Help Books
        An imprint of New Harbinger Publications, Inc.
        5674 Shattuck Avenue
        Oakland, CA 94609
        www.newharbinger.com

"A Person Just Like Me" adapted with permission from JOY ON DEMAND by Chade-Meng Tan. Copyright © 2016 by Chade-Meng Tan. Reprinted by permission of HarperCollins Publishers.

"Soles of the Feet" adapted from Singh, N. N., Wahler, R. G., Adkins, A. D., & Myers, R. E. (2003). Soles of the Feet: A mindfulness-based self-control intervention for aggression by an individual with mild mental retardation and mental illness. *Research in Developmental Disabilities*, 24(3), 158–169. Copyright © 2003 Elsevier. Adapted with permission.

Illustrations on pages 10, 11, 28, 81, and 82 by Kate Murphy

Illustrations on pages 7, 14, 23, 30, 34, 42, 50, 53, 64, 73, 88, 92, 97, 100, 101, 124, 130, and 135 by Zanne deJanvier

Cover design by Amy Shoup; Acquired by Tesilya Hanauer; Edited by Jean Blomquist

All Rights Reserved

Library of Congress Cataloging-in-Publication Data on file

20   19   18

10   9   8   7   6   5   4   3

For the young people in my life:

India, Mackenzie, Ethan

Evan, Dessa, Hailey, Makenzie, and Ethan

And for all the teens that I have taught in my classes,
and all the teens beyond my classes:

May you always feel completely safe and loved,
and accept yourself as you are.

contents

foreword

Let's face it—adolescence can be an awkward and often painful time of life. The incredible advances that happen at this stage, such as increased ability to monitor one's own thinking (what's known as metacognitive skills), can lead to newfound concerns that didn't plague us as much during childhood. What kind of person am I? Am I a good person or a bad person? What do people think of me? The process of identity formation, which is the central task of adolescence, can also bring a lot of stress if we're worried about whether or not we're good enough. Also, because teens don't yet have a lot of experience in the world, they can have a strong sense that their thoughts and feelings aren't shared by other people. This means that being a teen can be pretty lonely, even when a teen is surrounded by friends and family.

This is why self-compassion is a life-saver for adolescents. Self-compassion basically means being a good friend to ourselves when we struggle. When we're self-compassionate, we are kind rather than harsh with ourselves, remembering that imperfection is part of being human; and we're aware of what we're experiencing rather than being lost in a storyline about it.

There are now hundreds of research studies that show that self-compassion is strongly related to well-being in adults. Its benefits include less anxiety, depression, and stress; greater happiness and life satisfaction; better coping skills; and better relationships. We also know that contrary to common fears, self-compassion increases motivation to pursue valued goals because it puts us in the supportive emotional frame of mind needed to do our best, while reducing fear of failure and performance anxiety. Put simply, being kind and supportive to ourselves, especially when faced with difficult emotions, helps us thrive and reach our full potential.

Karen Bluth, the author of this workbook, is the world's leading researcher on the impact of self-compassion for adolescents. Her research as well as that of others shows that teens who are more self-compassionate do better. They are less anxious, stressed, and depressed and more satisfied with their lives than teens who are harsh with themselves. And we know from the research that self-compassion is a skill that can be taught, helping people of all ages lead happier, more satisfying lives. So Dr. Bluth, along with her colleague Lorraine Hobbs, has also developed an eight-week program designed to teach the skill of self-compassion to adolescents called Making Friends with Yourself. This program is an adaptation of the self-compassion program my colleague Chris Germer and I created for adults called Mindful Self-Compassion.

This workbook, which contains a wonderful variety of practices and exercises drawn largely from the Making Friends with Yourself course, is the key that will unlock the door to self-compassion. It makes the process of developing self-compassion simple. Full of fun and meaningful activities, this workbook will help teens learn how to be kinder, more supportive friends to themselves as they navigate the turbulent waters of adolescence. Life will still be a challenge, but the skill of self-compassion can help you meet that challenge with courage, strength, and kindness. I wish this book was around when I was a teenager!

> —Kristin Neff, PhD
> Associate professor, University of Texas at Austin
> Author of *Self-Compassion: The Proven Power of Being Kind to Yourself*

introduction

Being a teen can be pretty tough. My guess is that you've probably noticed that. There are probably times that you really feel like an outsider, and then a minute later you might get a text from a friend and that feeling is completely gone. Sometimes you might cry for no apparent reason. Sometimes you feel alone, like you're the only one on the planet who feels the way you do—and *no one* understands. And then you're with a group of friends, and you forget all those awful feelings. But then your friends leave, and there it is again—the loneliness, the sadness, the feeling that somehow you just don't measure up, that you're just not good enough.

Sound familiar? You're not alone.

Feeling this way is part of being a teen, and all teens experience these feelings at some time or another, probably a lot more than you think. It's just that we humans— teens and adults, too—get pretty good at hiding our feelings and putting up a good appearance, so it seems like everyone else is sailing along beautifully and we're the only ones who feel insecure and inadequate.

But trust me, those popular kids out there? The ones who seem like they have tons of friends, get great grades, win awards at sports, and do all this with no effort? They, too, feel desperately insecure at times.

I'm here to tell you that feeling like you don't measure up is a normal part of adolescence, and—as I'm going to explain in this book—you don't have to just suffer with it. There are things that you can do so that you don't have to endure years of your life being miserable, because that, certainly, is no fun.

What I'll show you in this book won't necessarily make your friends like you more. It won't get you into the college of your choice, or make your teachers give you less homework (although I'd love to be able to do that), or get you a date with the guy or girl of your fantasy. But it will help you like yourself more. It will help you be less harsh and critical of yourself—to be kinder to yourself. This will help you cope with these difficult things in your life—the stressors that all teens face in one way or another, the feelings of being imperfect, unworthy, and less-than that all human beings deal with—with greater ease. And being kinder to yourself will help you meet your goals (so you actually just might get into that first-choice college after all).

I bet I know what you're thinking—*But I don't deserve it. Some days I just say the stupidest things. Or do the dumbest things. And let's face it, I'm not as smart as others and I'm certainly not as attractive. Or popular. And besides, I'm fat*—or some combination of those phrases. And I bet you're thinking that others have way more going for them than you do. And that if you're nice to yourself, you'll turn into a lazy slob, do a lousy job on your schoolwork, get terrible grades, and never get into college, right? And then you'll not be able to get a decent job, and end up on the street eating out of garbage cans.

How do I know this is what you're thinking? Because this is how most teens feel. Most teens—if not all—are super self-judgmental. It's a part of the territory of being a teen. This book is going to explain why this is so, and more importantly, what you can do about it.

You don't have to struggle with self-criticism, self-loathing, or worse. This book will teach you about self-compassion and how you can be kinder to yourself and learn to like—even love—yourself *exactly as you are right now*, especially when you're feeling down. Self-compassion is about treating yourself with the same kindness and care that you give your good friends, especially when you're going through a hard time. In this book, you'll learn specific self-compassion exercises and meditations that will help you begin to do this. And you'll learn ways to use these tools throughout your day, as part of your everyday life. And we're going to have fun, too. We're going to do some art activities, a couple of music meditations, and a few creative projects.

My suggestion is that you take your time going through this book. Unlike a lot of things you learn at school, with this, it's good to be a slow learner. Take your time. Read, digest, do an exercise or two, and then maybe spend some time reflecting on what you got out of it. You're welcome to choose which activities you'll do, going for the ones that seem most helpful and passing over any exercises that don't. But I think you'll find most of them will help you be kinder and gentler with yourself, and help free you from that nagging, insistent Inner Critic that is so often standing over your shoulder and whispering in your ear.

One thing I want to mention. You'll notice that we go back and forth between talking to different genders in this book. That's because we want to include everyone—males, females, and teens who are questioning their gender. These tools are appropriate for everyone, so feel free to change the pronouns to suit you.

Many of the exercises in this book come from a mindful self-compassion course that I helped create and that I teach, called Making Friends with Yourself. Throughout the book, you'll see quotes from teens. These are real quotes from real teens who have taken my classes. Here's the first one, which a teen said at the end of the class:

What Teens Say

It made me less insecure and able to be myself more and kind of open up to others at school.

The experience this teen had is one that's there for you, too. Remember that you're not the only one dealing with being hard on yourself. We're all in this crazy world together. My suggestion to you in approaching this book is to be open. See what's there. Try it out. Make each activity an experiment, and see if you can have some fun with it. Play around with it, and see what works for you. Make it your own. You're not making any commitments here. You're just exploring a possibility—a possibility of seeing yourself in a new way, a possibility of being your own best friend.

Part 1

The Real Reasons That Teens Struggle—and What Can Help

chapter 1

changes, changes, and more changes

You wake up one morning, roll out of bed, look in the mirror, and notice that you look different. You're not sure what it is exactly, but it's maybe something about the shape of your face or the size of your hands. And more than that, you feel different. You seem to get upset easily, and sometimes cry for no reason, and then in the next minute you're as happy as a clam. What's going on? You used to just take things as they came, not getting too riled up about anything. But now it feels like you're on an emotional roller coaster half the time. There's a part of you that keeps criticizing you, judging your every move—there's this voice inside you that keeps saying, *You're not quite good enough* or *Who do you think you are, anyway? You just don't have what it takes!* And these changes seem to have happened overnight.

You get dressed—and what to wear? What are those cool kids who hang out by the lockers at school wearing these days? You look in the mirror again, and you look—well, not good. The scarf just isn't falling right. You change again. Uh oh, running late—no time for breakfast, carpool is honking, and you're out the door! Science class is first period and it's killing you. I mean, really, who can dissect worms first period, especially on an empty stomach? So you spend a lot of time staring at the ceiling or chatting with your cute lab partner. But your grade is suffering. And you keep hearing your teachers and parents saying, "You've got to pull that grade up in order to get into a good college!" Errgghh. The pressure. You hear that little voice inside you saying, *Come on, you know you can't do this. You're just not smart enough.* You feel like such a loser. And you feel like a loser after school, too, when you have

basketball practice and you practically trip over your own feet. (*You're so clumsy!* you tell yourself.) And then you feel like a loser again at your piano lesson, which doesn't go well because you didn't practice as much as you know you should have. (*Ugh. I know I should have practiced more this week—I really meant to. It's so embarrassing being on the same scale for three weeks in a row!*) And then finally, you're home. You're wiped out, feel like a failure, and have hours of homework ahead of you. You just want to crawl into bed. Or go back to elementary school when it seemed school and life were such a breeze.

What's going on?

changes are happening all around you

For starters, there are tons of changes going on in your *outside* world, in the environment around you. You've probably started a new middle or high school, and that's a big deal. It's super hard to start a new school where you don't know a lot of people. And chances are your relationships with your friends are changing—maybe kids you've been friends with since you were a toddler don't seem all that interesting anymore. And speaking of friends, you've noticed that you're spending a lot more time with them than with your family—and talk about "uninteresting," your family seems downright boring. And your parents don't seem to get you at all. They're way overprotective. (If this sounds like you, you might want to check out the movie *The Croods*. The star character, Eep, is a teen fed up with her family, especially her dad, who's all about keeping her safe and "guarded" in the family cave. You might relate to her desire to get out and explore the world—and the conflicts she and her dad get into.)

And then there's the workload—so much more homework and pressure than there was a few years ago! And soon you'll be making this major decision about college, and that seems overwhelmingly scary. On top of that, maybe you've started a part-time job. It's fun having some of your own money, but there doesn't seem to be time for much of anything else these days—let alone having fun.

What other changes have being going on in your outside world over the last couple of years?

_____

_____

_____

_____

Do any of these changes include you feeling self-critical in a way that you weren't before?

_____

_____

_____

_____

Let's be honest: all these changes are a lot to deal with, so don't be so hard on yourself! I know it doesn't feel good when you get into conflicts with your family, and I bet you end up criticizing yourself when it happens, but it just means you're all growing and stretching. And you know that desire you have to "fit in"? It's not a bad thing. It doesn't mean you're weak or a follower. We all need other people to survive—it's a basic human instinct to "belong" and feel part of group. And those cliques in school? They exist because everyone wants to feel like they "belong." If you've ever criticized yourself or felt bad because you've been rejected from a group, have no fear, you're not alone. It happens to many, if not all, of us. As for all those pressures and life stresses in your school and even your afterschool activities, if they ever make you feel inadequate, that's totally normal. In this book, you're going to learn some ways to be kind to yourself—to be self-compassionate—when these kinds of things happen.

One way to be kind to yourself in these moments is to do something that calms you and relaxes you a bit. Let's do an art activity that will help you slow down so you can really notice how you're feeling, which is a part of what we call *mindfulness*. Slowing down and noticing in this way is a first step toward being kinder to yourself. You'll need a pen or pencil for this activity.

try this: Mindful Art—Working on the Outside

Before you begin, make sure you're sitting in a comfortable position. Take a couple of relaxing breaths. After a minute or two, bring your attention to what's around you by listening to sounds. Notice sounds that are near and sounds that are far away. Take a full two minutes to listen in this way.

When you are ready to begin drawing, pick up your pencil or pen a little more slowly than you normally would. Let yourself be aware of the feeling of holding the instrument in your hand. Notice its temperature and texture. Pay attention to how you are gripping this object.

Then shift your attention to the drawing below. The mindful image below is divided into two sections: the outer border and the inner centerpiece. You will be working on the outer border first. Later you'll come back to the inner part.

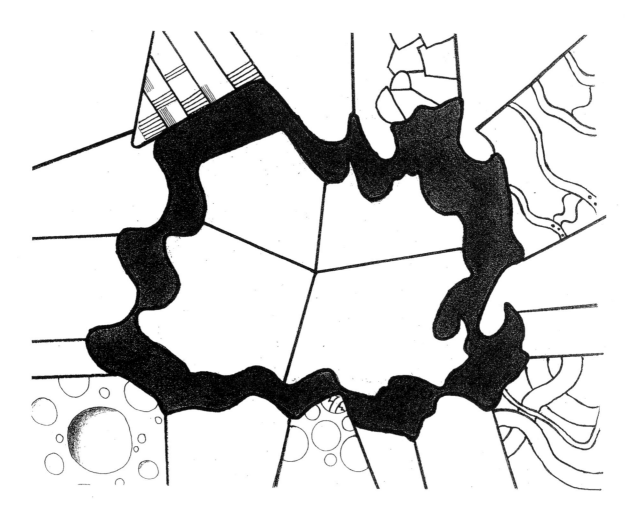

Let yourself examine the partial patterns. This is where you will begin to draw mindfully.

Some of the sections in the border have patterns started. Begin by completing these patterns. You can take a look at the sample patterns below as a guide. After you complete the patterns that have been started for you, fill in the other sections of the border with patterns that you create. Remember, for now, you're just doing the outer part. Take your time. This is not about creating a work of art; it's about noticing lines, feeling the pen or pencil in your hand, noticing the texture of the paper, and even hearing the sound of your pen or pencil on the paper. Go slowly, and draw the lines carefully. This should take you at least twenty minutes, and it might take a lot longer than that.

   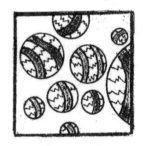

   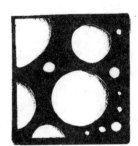

What was this like for you to draw in this way? (Circle any that apply.)

I liked it!

It made me feel calm.

I was anxious to get done.

It reminded me of coloring.

What's the deal with the inside??

Can we do more?

I was worried about it looking good, even though you told me not to.

I'm going to go buy a mindful coloring book now.

You may have noticed that you feel a bit calmer, a bit more centered or relaxed after doing this art activity. This is because you were focused on physical sensations—the feeling of the pen or pencil and the texture of the paper. When your mind is focused on physical sensations, you are in the present moment, and you're not dwelling on worries about the future or the past. This is what mindfulness practice trains us to do.

changes inside you

We've looked at what's going on in your *outside* world, but what's going on *inside* your body?

You've heard about puberty and hormones, I'm sure. You've heard about how your body is changing and getting ready for "reproduction" (ew). But there's a lot more to the story than that.

Your brain is also changing. Did you know that? Your brain is changing more than at any other time in your life except for when you were a toddler. Like the rest of your body, your brain is getting ready to take on life as an adult.

The most fascinating thing that is going on is that two separate parts of your brain are developing at the same time. One's called the *limbic system*, and it's the part of your brain that gets activated when you're feeling emotional—like when you're feeling afraid or angry. Because it starts changing when you're around eleven or twelve, that's when you probably notice that you cry more easily, or go from feeling chill to getting super angry in about a second and a half. It also makes you more prone to things like self-criticism and depression. And on top of that, these brain changes also make you more interested in taking risks and trying new things that you've never done before and maybe never thought you'd do. You know why? Because of these brain changes, you are less sensitive to getting really excited about things, and so you have to up the ante to get the same level of excitement than you did when you were younger. Weird, huh?

There's another part of your brain that changes as well—it's called the *prefrontal cortex*. (I know, all these fancy scientific names…) This part is responsible for "higher-level thinking," like planning, making decisions, weighing pros and cons of situations, and controlling urges. Up until you're about twenty-five, your prefrontal cortex is going through *pruning*, which means all the parts (also known as *neurons*) that are not being used check out and get eliminated. The parts that are used more often get stronger and can send signals to where they need to go more quickly.

Here's the wild part: both the limbic system and the prefrontal cortex start changing around the same time, but the limbic system—the part responsible for your emotions—changes much faster; it's pretty much done by the time you're fifteen or sixteen. The prefrontal cortex, the part that helps you think through things carefully and logically and make careful decisions is not finished developing until much later—around your mid-twenties or so. So scientists think that this might explain not only why teens sometimes do risky things without necessarily thinking about what might happen later, but also why teens feel such extremes of highs (when they experience exciting things) and lows (when they experience not-so-exciting or painful things). It may also explain why teens sometimes have a hard time seeing that the lows—like when you feel super self-critical and insecure—are not, in fact, here to stay. It's all because of what's going on in your brain!

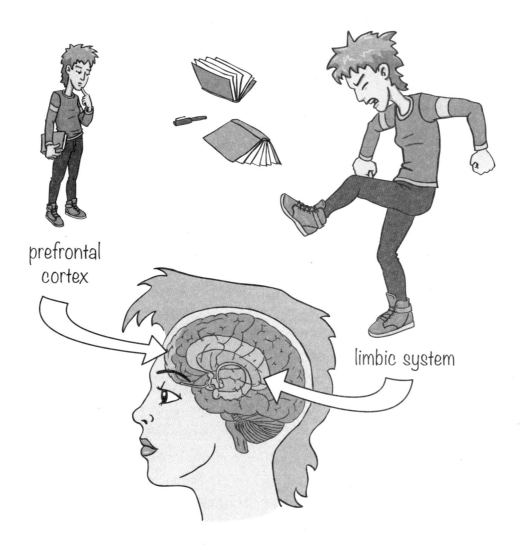

prefrontal
cortex

limbic system

(If you want to know more about all the changes going on in the teen brain and how these changes prepare you, in the end, to take on the challenges of the adult world, check out *Brainstorm: The Power and Purpose of the Teenage Brain* by Dr. Dan Siegel. It was written for both teens and adults, so you can share it with your parents, too.)

So now that you've heard about all these changes going on in your brain, which might be a bit unnerving, it's time to let go and relax a little, and dive back into our mindful art. Let's continue the mindful patterns that you began in the first half of this chapter.

try this: Mindful Art—Working on the Inside

Again, take a few quiet moments to bring your attention into the room. Notice any physical sensations that you might be feeling in your body. Maybe notice the contact of your body with the chair or couch or floor where you are sitting. Take a full minute or two to do this, and notice any sensations that are going on in your body right now.

When you are ready to begin, slowly pick up your pencil or pen. Let yourself be aware of the feel, temperature, and texture of the pencil or pen. Pay attention to how your hand feels on the object.

Taking your time, shift your attention to the drawing on page 10.

This time, you'll be working on the inner part of the drawing. You can use the patterns that you've started in the border or make up your own patterns for the inner part. Once you've finished that, you can add color to them if you'd like with either markers or colored pencils.

Remember, this is not about producing a great work of art, or even something you want to hang on your wall. It's about noticing what's happening *as* it's happening, noticing what you're feeling in your body in the moment when you're feeling it.

Take your time with drawing and plan on spending at least twenty minutes or so. It's not about finishing, but about noticing sensations.

How do you feel now? (Circle any that fit.)

Relaxed.

Chill.

The same.

Actually, this was kind of cool.

I never thought I liked art, but this wasn't bad.

Not my thing.

I like the color that I added.

Need a sharper point on my pencil.

So now that you've completed your second mindful drawing exercise, my guess is that you feel a bit better, right? So being mindful is a way that we can take care of ourselves, a way that we can be kind to ourselves. We can take time away from all the craziness of our day and just sit and draw. Or color. And that, my friend, is an act of self-compassion.

my thoughts

Use the space below to write or draw what you're thinking after reading this chapter.

Some ideas:

- Changes you've noticed in your external world—your school and relationships with family and friends

- Changes you've noticed in your internal world—your body and mind

- Changes you've noticed about how you feel about yourself, and how this might have changed over the last couple of years

conclusion

So many changes and transitions are going on in the teen years, both in the outside world—in school, in relationships with friends and with family—and inside our bodies, that it can feel overwhelming. Teens often feel like they don't "measure up" or aren't "good enough," because it's hard to adapt to all these changes—it takes time. So if you feel insecure and self-critical, know that you're not alone—this is part of being a teen and adjusting to all these changes. Read on to see how self-compassion can help by giving you tools that teach you how to be less hard on yourself and to accept yourself for exactly who you are!

chapter 2

what is self-compassion?

That's a pretty big question, isn't it? You've already gotten some hints about it in the last chapter. But rather than just coming out and telling you now, I'm going to allow you to discover for yourself what self-compassion is—and what it can do for you— by doing an exercise.

try this: Discovering Self-Compassion

Think of a time when a good friend told you something really bad that just happened to her. Maybe she failed a test, missed an important soccer goal, or the guy or girl that she liked asked someone else out. At any rate, she was feeling bad—like she was pretty worthless. Write about this incident here:

_____

_____

_____

Now think about what you said to your friend. And also how you said it—in other words, your tone of voice. Write it here:

_____

_____

Now, think of a time when *you* felt pretty bad—when something happened to *you* that made you feel like a giant idiot, like you didn't belong, or you were worthless, or like you just wanted to crawl into a hole and stay there forever. Maybe you failed a test, or missed an important soccer goal, or the guy or girl that you liked asked someone else out.

Take a minute to think about this.

Okay, now write about this incident here:

_____

_____

_____

Think back to this incident: what were the words that you said to yourself? Write down the words and describe the tone of voice that you used:

_____

_____

_____

What did you find out? Did you treat yourself the same way that you treated your friend? (Circle one.)

      Yes   No

My guess is that the answer is no! I'm right, aren't I? How did I know? Well, because of this simple fact: 78 percent of people are kinder to others than to themselves. So if you were one of those people that circled "No" because you were nicer to your friend than yourself,

Don't worry!

You are not alone!

In fact, most people are like you. They're kind to the people they know, but they judge and criticize themselves harshly. Why? you ask. Well, we don't really know. It might be that this is the way we're raised—that it's a part of our culture. We're told from early childhood that it's important to be nice to others, but that it's "selfish" or "self-serving" to be nice to ourselves. And on top of that, we're told that we won't be successful or happy in life unless we really push ourselves—we have to at least measure up to others, if not do better than them. And that often means being self-critical.

I bet you're thinking that if you're nice to yourself that you'll end up just lying around all day bingeing on Netflix TV series and YouTube videos about baby animals, right? That you'll never do homework, and then won't get good grades, then won't get into a good college, and you'll end up just being a major lounge lizard, right? Serious couch potato material?

Aha, but this is wrong! We know from research that folks who are kinder to themselves are actually *more* likely to be motivated to get stuff done. They are less likely to procrastinate, and more likely to try new things. You know why? Because they give it their all without worrying about failure or giving in to doubt. They know that if they don't achieve their goal, they're not going to beat themselves up; they'll just either let it go or keep trying. And they are overall happier as a result.

I know, it's not what we've been told. But stick with me—you promised you'd have an open mind, right?

We can be nice to others *and* be nice to ourselves, even at the same time! It all comes down to what self-compassion is.

definition of self-compassion

Treating ourselves, when we're going through a hard time, the way we would treat a good friend. That's it! We know how to be kind because we're usually really kind to our friends. So all we have to do is turn that kindness to ourselves. Sounds easy, right?

Actually, it's not hard. But it does take some practice, simply because we're not used to it. So if the practices in this book feel awkward or weird at first, don't worry. Just remember that it's just because they are totally new to us. And anytime we try anything new, it feels a little weird, right?

the three parts of self-compassion

There are actually three parts to self-compassion. It's helpful to know about them because all three parts can be used in practices both separately and together to help us be kinder to ourselves and not be so self-judgmental.

Self-kindness, or being nice to yourself, means saying kind words to yourself when you're having a hard time or finding yourself being self-critical—saying something like you would say to your friend. Maybe something like *I know how you feel* or maybe even *You rock!* It could also mean doing something nice for yourself—like watching a good movie or reading a fun book.

Mindfulness means paying attention to what you are experiencing in the moment with a sense of curiosity and without judgment. In the context of when we criticize or judge ourselves, paying attention to our experience in this way allows us to acknowledge the bad feelings we have, and to see that our bad feelings—our judgments, our self-criticism—are just that: feelings that will pass.

Common humanity is understanding that you're going through something right now that all teens go through. Hard to believe, I know, but I promise it's true! Other teens may not be going through the exact same situation, but all teens—and adults, too—feel upset, angry, hurt, sad, frustrated, lonely, and disappointed at times. And sometimes you feel these emotions all at once! This is part of being human— everyone feels these highs and lows, and teens especially. You might notice that sometimes you go from feeling really awesome to feeling really bad very quickly. This has to do with all the changes that are going on in your body during this time (we talked about this in the last chapter) and all those outside pressures—grades, school, sports, fitting in, all those things where you feel you might fall short or feel inadequate doing.

So try not to worry. You will get through it, as all of us do. Just remember that you are not alone, and feeling sad or hurt or depressed or angry does not mean you did anything wrong. It just means that you are human—and self-compassion can help.

Here's an exercise that'll help you actually *feel* what these things—self-criticism, mindfulness, common humanity, and self-kindness—are like.

try this: Hand Gestures

Stand up. Hold your two hands in front of you and make really tight fists. Squeeze as tightly as you can, and count to thirty—slowly. Think about how this feels. It hurts, right? Not comfortable, right? You want it to stop, right? This is what self-criticism feels like. It's tight, it hurts, it's really uncomfortable—it's not fun!

Now open your fists with your palms facing upward. What does this feel like? It's a relief, right? Feels open and relaxing, right? This is what mindfulness feels like. It's about feeling whatever is here in this moment, and letting it in.

Now stretch out your arms in front of you, keeping your hands open. What does this feel like?

Maybe like you're reaching out? Reaching for another person? About to give someone a hug? Maybe about to get a hug?

This is what common humanity feels like. Like reaching out and being connected to others. Being included. Being part of the group.

Now bring your hands in and put them over your heart. Allow them simply to rest there. Notice how it feels. Notice the feeling of the little bit of pressure or warmth on your chest. How does this feel?

_____

_____

_____

_____

_____

Maybe warm? Safe? Protected? Cared for? Loved? This is self-kindness, or self-compassion. Feels kind of nice, doesn't it?

So just putting your hands on your heart is an act of self-compassion. I bet there are other ways that you are already being kind to yourself. What kinds of things do you do when you are feeling tired, or disappointed, or hurt? How do you already make yourself feel better? Let's check this out.

try this: Self-Compassionate Things I Already Do

Here's a list of things that some teens do to make themselves feel better. You might want to get a blue pen and circle all the things that you're already doing. Then get a black pen and circle the things that you might want to try to do. With a red pen, cross out the things that you think you would never want to do. There's also a blank space at the end for you to add other things that I've left out.

Watch a funny movie.

Watch a scary movie.

Go for a bike ride.

Go for a run.

Play a game with your younger brother or sister.

Read *Harry Potter*. Or *The Hunger Games*. Or some other favorite book. Maybe for the fourteenth time!

Read a book that you would never read for school.

Curl up with the family cat.

Take the dog for a walk.

Play with the family guinea pig.

Call a friend.

Text a friend.

Go shopping and buy something nice for yourself.

Go outside with a sketchpad and sketch something.

Go outside and shoot some hoops.

Go outside with sidewalk chalk and draw something fun on the driveway.

Do a craft of your choice.

Put on your favorite music and dance like crazy.

Play a game (or two or three) on your phone.

Go for a hike in the woods.

Wash your face with a washcloth and cool water.

Make yourself a healthy snack.

Write in your journal.

If you don't have a journal, start one. All you need are a few blank pages.
Or a computer.

Take a nap.

Pet the family hamster.

Take a long, soothing bath.

Put on headphones and listen to your favorite music with your eyes closed.

_____

_____

_____

_____

_____

_____

_____

_____

Don't worry about having to think of every single thing right now. This is just a list to get you started. You'll be learning *lots* of different ways to be nice to yourself in this book, and you can always come back to this list to add things as you think of new ones.

Right now, how about trying one of these? When you do something nice for yourself, you feel better, and you might have more energy to do other things. Especially if you're mindful when you do it, and notice physical sensations—the feeling of the bathwater or the soft fur of your pet—you may really feel a bit rested afterward.

Write which one you're going to do:

_____

Go ahead and try the activity.

Write about how you feel now. You can use your own words or any of these: terrible, pretty good, not bad, so-so, feeling great today, happy as can be, depressed, sad, angry, frustrated, terrific, jumping for joy, the worst ever.

_____

What Teens Say

What do teens say when they've practiced self-compassion for a few months?

I feel more comfortable with myself. That's what I think.

It helps me feel more in my skin, I guess.

I guess I don't worry now so much about others liking me—because I like me!

What do you think it would be like if you made it a regular practice to be self-compassionate—and you did these self-compassionate things every day, not just once in a while? That's what we'll be learning and practicing in this book.

Mindfulness—paying attention with curiosity and without judgment—helps us be aware of how we treat ourselves so that we can be kinder to ourselves. This next art activity is a mindfulness activity—it helps us pay attention and teaches us to "let go" when we judge ourselves.

Remember the hand gestures that you did earlier? Here we're going to do a contour drawing of a hand—it will help us remember the hand gestures, and how mindfulness, self-compassion, common humanity, and self-kindness feel in our bodies.

try this: Mindful Hand Drawing

Remember, the mindful art activities in this book are not about creating a work of art to hang on your wall. They're about observing and taking your time to really notice your sensations—the feeling of the pencil in your hand, seeing the details of the lines as you draw and the spaces in between the lines, and even feeling the pencil on the paper. It's not about finishing the piece. In fact, don't worry about finishing—this isn't school.

You will need a pen or pencil and paper for this exercise.

When you're drawing, be aware of any judging thoughts coming up, and see if you can gently let those thoughts go. For example, judging thoughts might be things like *Eww, this looks sloppy* or *I bet of everyone who has this book, my drawing is the worst.* If you notice those thoughts arising, remember that this is totally normal and all of us have these thoughts. Remember that this exercise is about noticing sensations, and not about creating a finished work of art. In upcoming chapters, we'll learn how to deal directly with these judging thoughts. For now, it's just about drawing.

Begin by tracing an outline of your hand. Allow yourself to work slowly as you trace the outline so you can focus on the details you see on your own hand. After you trace the hand, begin to draw straight parallel lines on the background surrounding the hand outline. When the line reaches the outline of your hand begin to curve the line to reflect the contour of your hand. The line will become straight again when your pen or pencil returns to the background. Remember to go super slowly. There's no prize for finishing quickly.

Next, create mindful designs (like the ones in chapter 1) in the empty spaces. You can use the mindful designs from chapter 1, or you can create your own designs. The main point is to slow down and notice what you're drawing as you draw it, to notice sensations as you go.

If at any point you notice your mind wandering, gently guide your attention back to the sensations of holding the pen or pencil in your hand, or the sensation of your pen or pencil on the paper.

If you'd like, you can fill in the designs with color.

Here's an example drawing that my friend, Kate Murphy, drew:

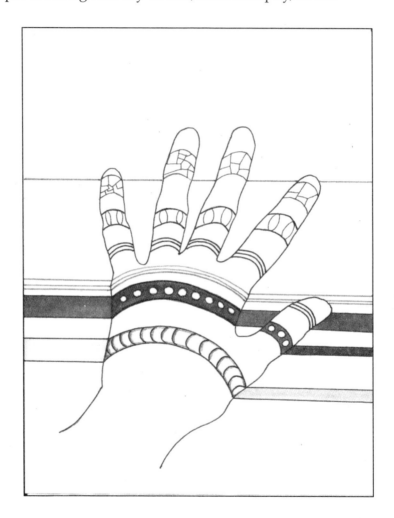

So remember, drawing mindfully in this way—taking your time, simply noticing your sensations as you're drawing, without judging yourself or your work—can be remarkably restful. And guess what? You can do it anytime!

my thoughts

Write or draw your response about anything you're thinking in reaction to this chapter. Here are some ideas you may want to respond to:

What would it be like to be kinder to myself and not beat myself up so much?

_____

_____

Is there anything that scares me about being kind to myself? If so, what is it?

_____

_____

Anything else you're thinking about?

_____

_____

_____

_____

conclusion

We're beginning our self-compassion adventure! And it *is* an adventure. It's opening a door into new, uncharted territory—territory that's exciting and that has the potential to offer us a new way of looking at ourselves, a way that makes us feel a whole lot better about ourselves, both in the times we're feeling okay and also when we're feeling low and self-critical. In the next chapter, we'll take a closer look at the first part of self-compassion—self-kindness. Are you ready? So strap yourself in, get ready, and here we go!

Part 2

Finding Self-Compassion

chapter 3

how do I begin to be kind to myself?

You're sitting in math class and the teacher is handing back a test you took last week. Your palms are sweating, because you don't think you did very well on it, but still you're hoping you did better than you thought. You could really use a decent grade on this test because your average this semester is—well, let's just say it's been better. The teacher is walking toward your desk, your heartbeat speeds up, her eyes meet yours for just a second, and then she places the paper face down on your desk. Instinctively you turn it over—your heart sinks. It's worse than you imagined. You feel so stupid. Ugh, there's that sinking feeling in the pit of your stomach, and the words *You're such an idiot* go through your mind again and again and just won't stop.

Or maybe you're at the lunch table in the school cafeteria, and you overhear a conversation—the guy that you could have sworn was about to ask you out just asked someone else to the school dance. Feels like you just got punched in the stomach. You try to swallow the bite that you just took of your pizza, but it seems to be caught in your throat, and you feel yourself fighting back tears. You say to yourself, *You should have known better. He was never going to ask you out—you're too ugly. Whatever made you think he liked you? What an imbecile you are! You are completely clueless.*

We've all faced scenarios like these that leave us feeling devastated. We have all experienced the horrible, intensely awful feelings—the self-judgment and the self-criticism—that go along with them. Unfortunately, being hurt, angry, alone, or sad is part of the experience of being human. So what can you do when something awful happens—when you feel like your world is closing in on you and you want to climb into a hole and not come about for about a hundred years?

First, know that there are things that can help so those experiences don't hurt so much. I'm going to tell you about two of them right now.

The first is called Soothing Touch. Simply put, it involves putting your hand on your heart, or stroking your cheek, or doing some other gesture that is soothing to you. It works because as mammals we are hardwired to be comforted in this way. As infants, we are soothed by our parents who rock or cradle us, just as all baby mammals are soothed by their mothers. Oxytocin, released in our bodies when we

receive or give a soothing touch, relaxes us and is even called the "feel good" chemical. If you have a dog or cat, think about how it is soothing to both you and your dog or cat when you pet it. When done in a gentle and caring manner, physical touch can be remarkably soothing.

So let's try out a few soothing touch gestures that will release oxytocin and relax you. You can use them the next time things get tough. If you practice them now, you'll get a sense of which one feels right for you and you can use it later. You can then use this gesture either by itself or along with other practices. As you're going through these gestures, notice how each one feels. It's best to do these slowly, taking five to ten seconds or so with each gesture so that you can really notice what each one feels like. After each gesture, rate how it feels to you on a scale from 1 to 10, with 1 being "eh, I could take it or leave it" and 10 being "super soothing." Put your rating on the line at the end of each gesture listed on the next page.

try this: Soothing Touch

1. Place one hand over your heart. Let it rest there for a bit. Notice the slight pressure of your hand, the warmth of it on your chest, and maybe how it seems to make your heart feel protected. _____

2. Place both hands over your heart, one on top of the other. Let them both rest there. Notice what that feels like. _____

3. Gently stroke your cheek with one of your hands. Notice that feeling. _____

4. Now cradle your face with both hands, allowing your face to rest in your hands. How does this feel to you? _____

5. Cross your arms and give yourself a little hug. You can also stroke one of your upper arms with the opposite hand. Notice what this feeling is like. _____

6. Hold one of your hands in the other hand in your lap. You can even stroke one of your hands with the other. This is a "handy" (no pun intended!) gesture because you can do it without anyone even knowing what you're doing. _____

7. Reach over and give yourself a pat on the back. How does that feel to you? _____

8. Make a fist with your right hand and put it over your heart. With your left hand, hold your right wrist. Notice how this feels. _____

Did you find a gesture that was particularly soothing to you? If so, write it here:

_____

What did it feel like when you did this gesture?

_____

_____

35

What Teens Say

I felt like having my hands across my chest was very anchoring. I really liked it; it was something different for me. It was nice. Very relaxing and comforting.

You can use this gesture by yourself when you're feeling upset or sad, or you can use it along with the second practice that I'll teach you now.

try this: A Moment for Me

You may read through this meditation below, or if you prefer, you can download the audio file for it at http://www.newharbinger.com/39843.

This practice is called A Moment for Me because it gives you a chance to take a moment to soothe yourself in the very moment when you most need it. This practice can be done either quickly in the moment or stretched out like a meditation. Let's try it as a meditation first.

Think of something that is a little bit upsetting to you, some situation that you are dealing with in your life right now. Because this is the first time you're doing this, it's best to think of a situation that is only mildly upsetting to you, and not the biggest thing that you're dealing with.

Get a good image of this situation in your mind. See if you can picture who it involves, what that person looks like, what the room or space looks like. If you'd like, you can either write about it or draw it here:

The next part has three steps.

1. Say these words to yourself, *This is a difficult moment*. Or maybe *This really sucks*. The exact words that you use aren't important—what's important is that you acknowledge that this moment, this experience of this particular event, is tough. It may seem obvious to you, but often we hide from our feelings, and bringing them fully into our awareness is super helpful in allowing us to work with them. How do you feel when you say this to yourself?

    _____

    _____

    _____

2. Next, say to yourself, *Feeling this hurt* [or anger, or frustration, or whatever it is] *is part of being human. We all feel these feelings at some time or another, and teens often feel them more deeply than others because of all the changes that we're going through. It's hard being a teen! And there is not a human being on this planet that has not felt these feelings. I am not alone.* Or you can say something similar to yourself that makes sense to you.

    Write your words here and write about how you feel after you say the words. Remember, however you feel is perfectly okay. There's no right or wrong way to feel.

    _____

    _____

    _____

3. And now you can use the soothing touch gesture that you found earlier that worked for you—maybe putting your hand on your heart, stroking your cheek, or patting yourself on the back. Now say to yourself, *May I be kind to myself in this moment*—asking yourself, *What words do I most need to hear that would make me feel better right now*? If those words come to you, you can write them here, and write about how you feel when you repeat these words to yourself:

    _____

    _____

    _____

Now try saying these words to yourself. For example, if the words that you might need to hear are "You belong," then say to yourself, *I belong.* If what you really need to hear is "You are loved," then whisper to yourself, *I am loved,* or you might even want to say to yourself, *I love you.* I know that, at first, it might feel weird talking to yourself in this way, but after you get over the weirdness of it, it might feel kind of good. Maybe even a few phrases have come to you. You can write them here:

_____

_____

_____

Sometimes people have trouble finding the right words to say to themselves. Don't worry if that's you—this happens a lot. It's simply because we're not used to being nice to ourselves. So try this: Ask yourself, *What would I say to a friend who was struggling with this same situation?* For example, suppose your situation is that you're feeling bad because you weren't very nice to your little sister and she really got upset and she started to cry. You can't think of the words that would make you feel better about this. Nothing comes to you. Now imagine that it was your friend who was mean to her younger sibling and felt bad about it. What would you say to your friend? Maybe something like "It's okay, you're human. It's okay to make a mistake. Just give your sister a hug, say you're sorry, and she'll get over it."

Write the words that you would say to a good friend who is struggling with the same thing that you are:

_____

_____

_____

Now say these words to yourself. For example, in the last situation, you might say to yourself, *It's okay, I'm human, and all humans make mistakes! I'm still a good person and a good sister!* What words would you say to yourself? Write them here. As you're writing, really think about the meaning behind these kind words and that, as a human being, you deserve kindness.

_____

_____

_____

How do you feel after doing this exercise? What was it like for you to say these kind words to yourself?

_____

_____

_____

_____

Is this something that you can see yourself doing in the moment when something difficult happens that makes you feel bad? (Circle one.)

Most definitely!!   Never!   Maybe

Still thinking about it   I'm going to see what else is in this book

What Teens Say

I've tried a Moment for Me a few times, like when I'm really stressed out about something. I just take a break and put it into perspective and say it's not really that big a deal.

So soothing touch, in combination with the practice A Moment for Me, can really comfort you when you're feeling bad. We often notice that putting a hand over your heart makes your heart feel protected and safe. And taking a moment to acknowledge that the pain is there, and remembering that this pain is part of life and something that we all experience, feels like a bit of a relief. Of course, the pain itself doesn't feel good, but knowing that it doesn't mean that it's your fault, or that you've done something wrong—that helps. And saying kind words to yourself helps, too. At any rate, when we stop trying to get rid of the bad feelings—because let's face it, that just doesn't work—and soothe ourselves just because we're feeling bad, the pain lessens a bit.

my thoughts

Feel free to draw or write about your reactions or responses to this chapter. What would it be like to do the practice A Moment for Me just when you experience something difficult or just when someone says something that hurts your feelings or upsets you in some way? What would it be like to be kind to yourself in the midst of this moment of feeling hurt?

conclusion

If you liked these practices, great! Use them whenever you notice that you're starting to feel bad. And remember that these practices aren't necessarily about getting rid of

the bad feelings—as much as we'd like to do that, it's just not possible. They're about being kind to ourselves right in the middle of the hurt—*because* we feel hurt. We're holding our tender and bruised heart gently, the way we might hold a newborn chick—with affection, warmth, and care.

In the next chapter, we'll get into a specific practice that really helps us both heal our hearts from the bruising of self-criticism and be kind to ourselves and others. This practice

has been around for thousands of years—and people are still doing it! More than that, recent research has shown that it builds our resilience—that is, it makes us more able to deal with difficult situations and bounce back from them.

chapter 4

opening my heart and letting kindness in

You're already kind to yourself in some ways, right? Like the ways that we talked about in chapter 2, such as taking a bath, playing with a pet, or listening to music. In this chapter, we're going to go one step further. We're going to learn ways that will make self-compassion more enduring so that it becomes something that you can take with you at all times, that you can call upon when you most need it.

How do we begin? My guess (because I teach a lot of teens and adults) is that you've had lots of times when you are pretty self-critical. Lots of times when you might even feel like you hate yourself. When you feel like no one likes you and you feel totally like an outsider.

So how can we turn this around?

In this chapter, we're going to learn a *formal practice*. The "formal" part means that we put time aside each day to do it. The word "practice" is used because we can do it over and over—we "practice" it, like soccer drills or playing scales on the piano. This formal practice works because it helps us cultivate kindness for ourselves in a very direct way—not just in the moment when we're struggling, but all the time, so it becomes part of who we are and starts to feel natural.

This practice is called "Lovingkindness." It's actually a very old practice that's been around for thousands of years. It's also called "friendliness." So it's about establishing a "friendly" attitude toward ourselves and others. This may sound pretty basic, but actually we often don't have a friendly attitude toward ourselves, do we? We often are pretty hard on ourselves.

Let's try this out and see how it goes. And then we'll talk about it later.

try this: Lovingkindness for Someone You Care About

If you can, please listen to the audio of this practice, which you can download from http://www
.newharbinger.com/39843. It will be easier to listen while you're doing the practice than to read it while
you're trying to do it.

1. Get into a relaxed position, either sitting or lying down. Make sure you're in a comfortable
   position where you'll be able to be for about twenty minutes.

2. Now think of a living being who makes you smile. This can be a friend, your grandparent,
   a favorite teacher, or even your cat or dog—any living being that makes you feel good
   when you think of them.

3. Now this living being, like all living beings, wants to be happy. So silently repeat the following phrases for this living being, seeing if you can connect to the feelings behind the
   words so that you're not just repeating words but really feeling the feeling associated with
   the words. Repeat these phrases slowly, so that you can really feel the meaning behind
   the words. Take your time. You may want to repeat these phrases several times.

   May you be happy. (Think about what it feels like to be happy.)

   May you feel loved. (What does it feel like to feel loved? To be loved just the way you are?)

   May you begin to accept yourself just as you are.

4. Repeat these phrases until you're really feeling a little bit of the wish for happiness, love,
   and self-acceptance for the other being.

5. In your mind, imagine that you are widening the circle that's surrounding this being so
   that you're including an image of yourself. In other words, you are now imagining yourself standing with this being that makes you smile—you're both standing there together.
   Now, taking your time, repeat the following phrases for both yourself and the other being.
   Remember to go slowly so that you can really feel the meaning, sense the feeling, behind
   the words:

   May we both feel happy (remembering the feeling of being happy).

   May we both feel loved (remembering the feeling of being completely and totally loved).

   May we both begin to accept ourselves just as we are (remembering this is just beginning
      to accept yourselves just as you are—simply being open to the possibility that you
      could accept yourselves exactly as you are, maybe at some time in the future).

6. Now, letting go of the image of the other being that makes you smile, letting the image fade gently into the background in your mind so that there's just an image of yourself in your mind, an image of yourself as you are today. Now very slowly, repeat these phrases just for you. You may want to spend more time repeating one phrase, to give it emphasis, feeling a bit of the wish of being happy and loved, and what it would be like to accept yourself just as you are. Take lots of time to do this. Remember, there's no prize for finishing quickly.

May I feel happy.

May I be loved.

May I begin to accept myself just as I am. (Just be open to this possibility.)

So what did you think? Or maybe I should say, What did you *feel*? Did it feel good to repeat the phrases for the being who makes you smile? Did it feel a little weird to say the phrases to yourself? Or did you maybe love the whole practice?

Here's the deal. There's no right or wrong way to feel. Whatever you feel is perfectly right for you. It helps to allow yourself to feel whatever it is that you're feeling. If you can find where the feelings are in your body, just notice those physical sensations. And just let them be there.

If you felt anything in your body as you did the practice, you can describe it here:

_____

_____

Now let's talk about this practice in parts.

Here's the first part: What was it like when you repeated phrases for the being that makes you smile?

_____

_____

And the second part: What was it like when you included yourself with the being that makes you smile?

_____

_____

And the third part: What was it like when you repeated the phrases just for you?

_____

_____

Remember, there are no right or wrong answers—whatever you feel is totally fine.

Did you feel like it was a little harder to repeat the phrases just for yourself?

_____

_____

So we're raised with this crazy notion that we're being selfish if we are nice to ourselves, even though we know that being nice to ourselves does not mean that we can't also be nice to others! On top of that, our culture and the media promote this idea that we're not good enough the way we are, and we have to buy a zillion products so that we can be "acceptable." As a result, many of us feel like we don't deserve to be happy, or loved, or safe. If that's you, don't sweat it—know that lots and lots of folks feel that way, and especially teens. In fact, there are a whole lot of folks that feel like they don't deserve to feel good about themselves. That's why I'm writing this book, in fact. Also, that's why we say phrases like "May I _begin to be open_ to the possibility of accepting myself just as I am."

Oh wait—did I not say that? Right. Brings me to another point. If these phrases aren't super meaningful to you, you can change them so that they _are_; you can personalize them. Here's how you do it.

try this: Finding the Right Phrases for Me

Ask yourself, *If I could, what words would I like to hear every day for the rest of my life? What words would I want to hear that would make me feel worthy and appreciated, valued and valuable—that every time I heard these words I would be so grateful and happy for hearing them? What words would I want whispered into my ear every day if I could have that happen?* Maybe something like "You are loved" or "You belong here." Maybe something else. Here are some words and phrases that you can use if you'd like:

Accept yourself

Appreciate myself

Know that I belong

Calm

I care for you

I'm here with you

You are worthy

Happy

Healthy

Honor yourself

Kindness

Know that you are loved

Peaceful

Safe

Remembering self-compassion

Support

If you'd like, write the phrases that resonate with you here:

_____

_____

Now make these words into a wish. Maybe "You are loved" changes into "May I feel loved" and "You belong here" becomes "May I feel like I belong."

You can write your "wish" phrases here. It's best to have at least two phrases, but fewer than four (just so they are easy to remember):

_____

_____

_____

Remember that we're just being *open* to feeling this way. Feeling comfortable with being kind and loving toward ourselves often takes time, and sometimes it happens a little bit at a time so you don't even notice what's happening. And then one day you turn around and you realize that you're not as hard on yourself as you used to be. That you like yourself a lot more. That you don't get so upset if someone says something that's not so nice—you realize that it might be their issue, actually.

So take your time learning to be kind to yourself. Remember it's totally okay to be a slow learner! This is not school! And by the way, even if you don't have a lot of time to do this practice, a little bit each day helps. You can do five minutes, right? Five or ten minutes a day adds up to a lot over time!

What Teens Say

I really like meditation specifically. I found that was really helpful, just to kind of relax a little bit from all the stress in regular life.

Here are some hints that will help you remember to practice lovingkindness in a formal way so you can continue to build kindness toward yourself. These suggestions will work for all the other formal practices in the book also.

1. Most importantly, remember that doing this practice teaches you to be kind to yourself. It's a time for you to remember that you are deserving of kindness, care, and love. It is *not* like homework or one of your chores! It is simply a time of remembering and generating love for yourself.

2. Set aside the same time each day to do a practice. That way it will become part of your daily routine, like brushing your teeth or getting dressed. Start with five minutes of practice, and then gradually build to doing about twenty minutes or so.

3. Do a practice in the same place each day, like a corner of your bedroom, on your bed, or in a favorite chair. That way that space will become a special "practice" place.

4. Set a timer on your phone to help remind you to practice.

5. Use the meditations associated with the practices. (You can download the meditations at http://www.newharbinger.com/39843.)

Personal Story from a Teacher

Sharon Salzberg, one of the best-known teachers of lovingkindness practice, tells a story about when she was first learning it. She was on a retreat, which means that she was doing this practice all day every day, like maybe eight hours a day or more—no, I'm not kidding! After a week or so, she didn't feel like anything was "happening." And then she had to leave the retreat for some reason. She was in her room, getting ready to leave, and knocked over a vase, which smashed to the floor. The first words that came to her mind were *What a klutz!* And then the second words that came to her were *But I love you anyway!* She says that she knows that those second words—"But I love you anyway"—would not have come to mind had she not been doing this practice.

my thoughts

Take some time to write or draw your reactions to this chapter.

For example, what would it be like if you spent time each day saying lovingkindness phrases to yourself? How do you imagine this might change how you feel during the day?

_____

_____

_____

_____

_____

conclusion

Learning how to be kind to ourselves is a process. It doesn't happen overnight, but it does happen eventually when we practice it. My guess is that, like most of us,

you've spent a lot of time being hard on yourself, and so turning that around and being kind to yourself may take some time. But the lovingkindness practices, especially when you make the phrases your own, help make that happen. And mindfulness, too—which we'll delve into in the next chapter—helps us be with whatever we're experiencing at the moment, rather than trying to push it away or make it different than it is. So are you ready to learn more about mindfulness? Check out the next chapter!

chapter 5

mindfulness, automatic pilot, and learning to pause

Did you ever walk into a room and wonder how you got there? Or have you maybe driven to school and you don't remember anything about the drive? When we do things on "automatic pilot" like this, we're not being mindful.

Remember, mindfulness is about paying attention to each moment with curiosity. It's about noticing what is happening as it's happening—noticing the sensations in our body, our feelings, and our thoughts in each moment. It's particularly important when we're talking about self-compassion because we have to be aware of how we're feeling before we can give ourselves the compassion that we need.

So enough of me talking about what mindfulness is—let's have an actual experience of mindfulness!

try this: Listening to Sounds

Right now, spend about one minute paying attention to sounds. If you have a timer or stopwatch on your phone, set it for one minute. Listen to the sounds around you—sounds close by and far away.

.... listening .... hearing ....

What did you notice?

_____

_____

Did you hear more than you usually do when you're not given the specific instructions to pay attention?

_____

So mindfulness is making a point to pay attention—making the *intention* to pay *attention*.

Did you notice your mind wandering at all when you were listening to sounds? Maybe you noticed thoughts arising in your mind? If so, this is totally natural—it's how the mind works.

Remember about being on automatic pilot and not remembering how you got from one place to another? This also happens when the mind wanders.

So why do our minds wander? And what happens when our minds wander?

Here's the story: We are hardwired to keep ourselves safe. From an evolutionary standpoint, the name of the game is to keep ourselves safe and alive so that our species will continue. That's what we're hardwired to do. Back in prehistoric times, this meant when our ancestors were out in the forests hunting, they would have to constantly watch out for anything that could hurt them—and it was usually some kind of predator.

Today, the kinds of things that hurt us, the things that we watch out for, generally aren't physical things, but emotional things—like someone hurting our feelings. But guess what? Because we're still wired to look out for things that can hurt us, we are constantly on the alert, constantly keeping an eye out for any potential problem. And when we find something that we think is going to harm us, we go into what's called the *fight or flight mode*.

what's the fight or flight mode?

Our bodies respond like this:

- We perceive something as dangerous → a message (in the form of a hormone) gets sent to our brain → which then sends a message (in the form of another hormone) to our pituitary gland at the base of our brain → which then sends a message (yes, another hormone) to our adrenal glands → which then release more hormones to get our body ready to fight or flee.

- Blood gets diverted from our digestive system because we don't need to be digesting food right now (it's not a top priority!) to our arms and legs so we can run or fight. This is why we get the feeling of "butterflies" in our stomach.

- Our pupils dilate so we can see more.

- Our heart starts pumping more (that's the feeling of our heart pounding) to get the blood to our arms and legs.

- We might start sweating more to keep ourselves cool when all this is going on.

And you know what? Whether the threat is real or we just *think* it's real doesn't make a difference—our bodies respond the same way. Here's a funny story that actually happened to me.

Personal Ridiculous Story from a Teacher (Me)

I was at the gym working out on an exercise bike. It was one of those bikes that has a screen in front of it. You choose the kind of terrain that you want to ride on, and there's a moving image of "your bike" as you ride along, with mile markers and a million other little things to keep you entertained as you're riding. Anyway, this day I was doing the "Alpine Slope"— "riding" on a windy mountain road with snow- covered peaks to the left of me and a treacherously steep drop-off to the right. I must have gotten distracted for a second, because when I looked back at the screen, I was headed straight off the road, about to careen fifty million feet down to my death! Or so I thought—for a second—until I remembered that I was on a stationary bike in the gym, safe as could be, with no sign of snow-covered peaks or mountain drop- offs anywhere. But my heart was pounding, my pulse was racing, and I had been genuinely terrified for a moment. And okay, I did get a good laugh out of it.

So whether the threat is real or not, we can still go into full fight or flight mode.

What does this mean for us?

We tend to worry about things that might hurt us (failing a test, doing something silly in front of others) so that we can be prepared if the worst happens—and we won't be caught off guard. This is what is called *negativity bias*. We tend to see the negative more than the positive. Because of our negativity bias, we have to make an extra effort to see things clearly, as they are, and mindfulness helps us do that.

Remember, mindfulness is about bringing our awareness to whatever is going on for us right in this moment, noticing what we are feeling, with interest and curiosity. So mindfulness helps us counter our negativity-bias tendency by bringing our awareness to what is *really* happening, not what we are afraid might happen.

And now for some shocking news—get ready for this:

When we're thinking, it feels like what we're thinking is absolutely, positively true, right?

But guess what?

Our thoughts are not facts!

This is true. Our thoughts—the things that come up in our minds and go through our heads—are not facts. I know they certainly feel like facts when we think them, but they aren't. They're often directed by our negativity bias and our fears. So, for example, if you think the self-critical thought *I am such an idiot because I totally said something really stupid at the party last night*, that doesn't mean that you *are* an idiot. That thought is probably based on a fear of looking stupid in front of others, and you thinking that thought is your negativity bias showing up. Why? Because deep inside you want to protect yourself from being hurt by people thinking you're an idiot. Get it?

My guess is that people aren't thinking about what you said at all—because they're worrying about what *they* said! At any rate, your thought about being an idiot or people thinking you're an idiot is just a thought—just something passing through your mind. It's not a fact at all! In fact, some people just call thoughts "mental secretions." (Ew!)

So mindfulness helps us see the facts, to see what's really here. It's about learning to *pause* and then remembering to pause before reacting—taking a breath or two when something happens that upsets you and noticing sensations in your body. It's about realizing when the fear-driven negativity-biased thoughts arise in your mind, and simply watching them as they fade away—just like clouds drifting across the sky.

By bringing our minds continually back to the present moment, as we do in mindfulness practices, we avoid taking the road to the fight or flight response. We stay present, and see the situation for what it is. So, for example, the pimple on our face that we think is this huge disgusting monstrosity that everyone is staring at and thinking that we're so ugly, is really just a pimple on our face. A slightly raised pink mark. That's it.

And you know what? Most of the time, no one even notices. Don't get me wrong— the fight or flight response is great to have when we actually need it—but most of the time, the fight or flight response is overkill for our daily worries. So, in other words, a pimple is just—well, a pimple.

And mindfulness practices can help us not freak out and not go down the fight or flight path. They help us stay with what is actually happening in the moment.

One simple way to practice mindfulness is to remember this: Physical sensations bring us to the present moment. When you're paying attention to your physical sensations, you're in the moment. And you know what? Your anxiety or worry cannot exist in the present moment.

As John Travis, a mindfulness meditation teacher, says, "Keep your mind in your body."

Here are two exercises that will help you do that, especially when you're in a situation where you're feeling inadequate or criticizing yourself—maybe about a test, a sports game you just played, or a situation with friends.

try this: Soles of the Feet

This practice is adapted from the work of Dr. Nirbhay Singh, a mindfulness researcher.

Please download the audio for this practice at http://www.newharbinger.com/39843. If you choose not to follow the recording, then do this practice slowly so that you can really notice what's going on in each moment. The whole practice should take at least five full minutes.

For this practice, it's best if you can take your shoes off.

1. Standing up, notice the feelings of your feet on the floor. Notice the whole bottom of the foot—the toes, the heel, and if the middle part of your foot, the instep, is touching the floor.

2. What does it feel like on the bottom of your feet? Does the floor feel hard, or is it cushiony? Does it feel cold or warm against your feet?

3. Now, very slowly, lean forward just a tiny bit. How does this change what you're feeling on the bottoms of your feet? Really take time to notice any changes in sensations.

4. Very slowly, lean back a tiny bit—just a fraction of an inch. Do you notice anything different happening on the bottoms of your feet?

5. Repeat this a few times, slowly rocking forward just a bit, and backward just a bit. What's going on with the bottoms of your feet?

6. Now lean to the right just a tiny bit. Again, notice any changes in sensations on the bottoms of your feet.

7. And now lean toward the left a tiny bit. Any changes going on down at the soles of your feet?

8. Repeat this slowly a few times—leaning just a bit to the right and then to the left, back and forth. What's happening with the bottoms of your feet?

9. Now make little circles with your knees, feeling the changing sensations in the soles of the feet.

10. Come back to center, keeping your attention on the soles of your feet.

11. You might take a minute to notice how amazing it is that the small surface area of the feet supports the entire body. Pretty wild, huh? And maybe take a moment to just appreciate your feet, feeling thankful for all the hard work they do all day long.

What stood out for you when you were doing this practice?

_____

_____

_____

When you were paying attention to the feelings on the bottoms of your feet, did you notice feeling any negative emotions—like sadness, anger, or hurt?

_____

_____

_____

Is this a practice that you can imagine doing sometime when you happen to be standing around anyway? (It's particularly helpful when you're starting to feel upset.)

_____

When we're paying attention to physical sensations like we did in this practice, our attention is in the present moment. Think of the senses—hearing, tasting, smelling, touching, and seeing—as doors that open to the present moment. When we go through those doors, we tend to *not* be in our heads, which means we're not dwelling in our thoughts—we're "letting go" of our thoughts.

The second mindfulness practice we'll do is a variation of Soles of the Feet. Instead of paying attention to our feet, though, we'll pay attention to feeling an object. This practice is a favorite of many teens.

try this: Here-and-Now Stone

Please download the audio for this exercise at http://www.newharbinger.com/39843.

This practice usually uses a small, polished stone. Any stone will do. In fact, teens have told me that you really don't need a stone at all—that sometimes they didn't have their stone with them and they did it with a bracelet, ring, pen or pencil, or any object that was nearby. So if you don't have a small polished stone, choose an object that can fit in the palm of your hand.

1. Study the object carefully. Notice all the colors in the object, the different hues, any lines or marks in the object. Notice how it reflects or doesn't reflect light.

2. Now notice the shape and the texture. Is it smooth? Rough? Maybe a little of each? Does it feel the same all over? Sometimes it's helpful to close your eyes when you do this—it's easier to get in touch with what you're feeling.

3. Does it feel cool to the touch? Warm?

4. Does it have a smell? If so, what does it smell like? Is it a pleasant smell or an unpleasant smell?

5. Really take time to get to know your object. Become so familiar with its characteristics that if it were in a pile of similar objects, you'd be able to find it in a second.

When you were studying your object, what thoughts were going through your head?

_____

_____

Did you have any upsetting thoughts? (Circle one.)

    Yes   No   I don't remember!

Remember that physical sensations are doors to the present moment. When we're focused on things like textures, smells, and physical appearance, it's easier to let go of troubling thoughts.

What Teens Say

I really liked the here-and-now stone. Yeah, especially during AP testing. I brought that with me and that was like phenomenal. It really helped.

I like the here-and-now stone because even if I don't do it with the actual stone, I can still do it with other things, like I can do it with my hand. I know that's kind of weird but...

I do that all the time at school. Even just walking from class to class, if you're stressed, you can pay attention to one specific thing, even if it's the tree or the wall. It distracts you from the other stressful things and you just focus on that—and it's fun.

ruminating and ruminating and ruminating and...

The practices I've guided you through also help us to stop spending time worrying about things that might happen in the future or dwelling on things that have already happened in the past—things that we can't do anything about anyway. This is called *ruminating*.

Ruminating also describes what a cow does with its cud—it chews it over and over. This is what we do when we keep thinking the same thoughts over and over, thoughts like *I'm too ugly! I'll never find a boyfriend!* or *I'm too dumb! I'll never get into college!* So ruminating is usually not very self-compassionate. And you know what? In research studies, ruminating has been linked with depression. People who tend to ruminate a lot also tend to be more depressed. That's because when we're ruminating, we're often thinking about all the things we did wrong and worrying about all the things that are bad that might happen to us in the future.

Mindfulness helps us *not* ruminate and stay more in the present moment. There's another mindfulness practice that you can do in the moment when you're noticing yourself ruminating or having upsetting thoughts. It's also a self-compassion practice, because it helps us soothe ourselves. This one is another favorite of teens—it involves music.

try this: Music Meditation

Please download the audio for this meditation at http://www.newharbinger.com/39843. The length of this meditation depends on how long your piece of music is. Your choice.

1. First, find a piece of music that is relaxing and doesn't have any words with it. It's important that the music not have words because words will get you to start thinking, and you want to stay out of your head and stay with your sense of hearing. Many teens really like music from *Theme from Silk Road* by Kitaro (https://www.youtube.com/watch?v=p5n57OSe8tw).

2. Next, get into a comfortable position, either lying down or sitting. Make sure you're in a place where you feel you can really relax. You can close your eyes, if you wish.

3. As you play the music, really pay attention. Listen to the tones, staying with the music, with all its movement, its ups and downs, its highs and lows.

4. When your mind drifts and you notice that you are thinking (which will happen at some point), just guide your attention back to the tones of the music.

5. Each time this happens—each time your mind drifts—just gently bring it back to listening to the music.

What stands out to you now after doing this practice? What was it like for you?

_____

_____

_____

Did you notice thoughts arising at some point? Were you able to guide your attention back to the music?

_____

_____

Being aware of when you are thinking, and developing the habit of bringing your attention back to the sounds of the music, takes time. And remember—mind wandering is not a bad thing! It's totally natural. Each time you notice it happening, just gently guide your attention back to the music.

What Teens Say

I definitely like the music a lot because that made me realize that I can do that every time I listen to music, and I always listen to music so that means that like, 24/7 I can be like, calm.

Another way we react to things that bother us—besides ruminating on them—is pushing them away. This is called *resistance.* Resistance works against mindfulness because mindfulness allows things to be as they are and resistance is wanting things to be different than they are.

Resistance is judging ourselves and our experience. It's that little voice in the back of our heads saying, *Who am I kidding? I am just not good enough to hang around with those kids!* or *Aaagh! I did awful at the game! I should have made that goal! I let my team down! I feel like an idiot!*

So what exactly does this mean? For example, imagine that you are in a play that you have been rehearsing for months. You're not the lead, but you have a pretty decent-size role. And others are dependent on you, because you're part of the cast. And there's all this excitement leading up to opening night—and then the curtain rises. You're onstage somewhere in Act I, Scene 2, you turn to the lead who is there onstage with you, open your mouth to say your lines, and nothing comes out! You've completely forgotten your lines! And the lead is looking at you, with this look of sheer panic on his face. And you just kind of stutter…

Later, you beat yourself up for forgetting your lines. You feel like you let the whole cast down! And as much as others keep telling you that it's okay, that this happens a lot, you feel like a total loser. Resistance is beating yourself up, wanting to change something that already happened—something that, let's face it, you can't do anything about. So *not* resisting would be just accepting the fact that what happened is over, and there's no going back.

So much of the time, we resist our lives being the way they are. We want our teachers or parents to be different, we want ourselves to be different, maybe we want our friends to treat us differently, or at the very least we want to have less homework. Letting these things be and accepting the way they are doesn't mean they won't change. They might change. But resisting the way they *are* will only cause more unhappiness for us. And it doesn't mean that we can't do something to help them change—we can. But in the moment when we want things to be different from the way they are, when we want to go back in time and not forget our lines, the resistance actually makes us feel much worse.

So—what we resist persists!

And when it comes to bad feelings, like anger or feeling worthless, resisting actually makes the feeling—the anger or the feeling of worthlessness—stick around longer.

the ninety-second rule

What we know from how the brain works is that when you feel a feeling—any feeling, good or bad—the brain chemistry responsible for that feeling stays in your body for ninety seconds. That's it! From the time that you start feeling angry to the time the "anger chemicals" are out of your body is just ninety seconds. Isn't that crazy? We all stay angry for way longer than ninety seconds!

Why does this happen? Why do we stay angry (or hurt, or sad, or frustrated) for longer than ninety seconds?

Chances are you think the thought again: *She really made me angry! She shouldn't have said that to me! She had no right! That was mean!* And the release of the anger chemicals happens again, and so you feel angry all over again for another ninety seconds.

And as long as you keep revisiting the thought, you'll stay angry.

For maybe a long time.

So how can mindfulness help?

When practicing mindfulness, each time the thought arises in our minds, we "let go" of the thought and return to an object of our attention. *Letting go* just means imagining that we are watching the thought float away, kind of like watching a cloud as it drifts along in the sky and out of our field of vision. Often, we use our breathing as an object of our attention, as an anchor to the present moment. The breath is often used because it's something that we always have with us—and we don't have to look for it.

In the next practice, we are going to use our breath to practice mindfulness. This practice is a core mindfulness practice that you can set aside time to do each day.

Before we start, let's see how you're feeling. In the circle below, shade in the amount of "good" you feel. In other words, if you feel really terrible, none of it would be filled in. If you feel great, all of it would be filled in. And, by the way, you get to interpret what it means to feel "good."

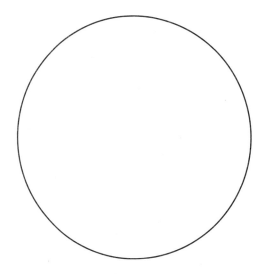

try this: Mindful Breathing

If you'd like, you can download the audio for this meditation at http://www.newharbinger.com/39843. Like all meditations, it's important to do this slowly, taking lots of time to feel your breath. The whole practice should take you at least ten or fifteen minutes.

1. Sit comfortably on a cushion or chair and, most importantly, find a position that will help you be aware and alert. It's best to sit upright rather than slouching, because this will help you stay alert. Alternatively, you can lie down on a couch or floor. (Just be careful that you don't go to sleep!)

2. You can do this practice with either your eyes opened or closed. If you choose to keep your eyes open, choose a spot about four feet in front of you on the floor to rest your gaze. You're not staring, but just resting your eyes on that spot on the floor with a "soft gaze."

3. Take a few slow breaths—notice the breath where it is most obvious to you. This could be at the tip of your nose as you're breathing in, or your mouth as you're breathing out. It might be in your nostrils as the air passes through, or in the movement of your chest or diaphragm area. (If you're lying down, you can rest your hands on your diaphragm, the area just below your rib cage. This way you can actually feel your diaphragm moving up and down with each breath. (Take your time to find the place.)

4. See if you can feel the breath from the very beginning of the in-breath all the way through to the end of the out-breath. Notice the movement of the breath, the temperature of the breath, and even the texture of the breath.

5. When you notice that your mind has wandered, which it will, gently bring your attention back to the breath. Don't worry about how many times you do this: your mind naturally wanders—this is its job. So when you notice that it's wandered, just gently—and without judging yourself—bring your attention back to your breath.

6. Feel your whole body breathe, gently moving with the rising and falling of the breath, kind of like the movement of the sea.

7. Allow yourself to enjoy the relaxing qualities of the breath, just like you might enjoy the feeling of a warm blanket. For the next few minutes, continue doing this practice.

8. Feeling your breath, noticing when your mind wanders, and coming back to your breath.

9. After a few minutes, releasing the breath, and allowing everything that comes to your awareness to be just as it is, just for now.

10. Take a moment to notice how you're feeling right now.

11. Slowly and gently open your eyes.

Now let's see how you're feeling. Shade in the amount of "good" you feel.

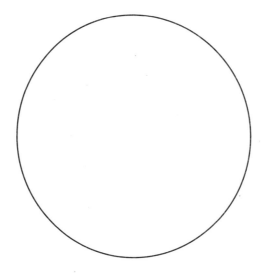

Trish Broderick, author of a mindfulness program for teens called Learning to BREATHE, suggests a mini three-breath version: just take three mindful breaths three times a day no matter what, even when you're not feeling stressed.

Easy to remember: three mindful breaths, three times a day. Many teens have told me that this practice really works for them.

You can use this mini-version of the Mindful Breathing practice right when you notice self-critical thoughts coming up. To try the three mindful breaths practice, please download the audio file from http://www.newharbinger.com/39843.

What Teens Say

The three breaths—I always use it when I'm in a situation when I need to do something. Like I'm getting really angry or something, I kind of just take three breaths and I really think about them and it kind of calms me down.

I started having a panic attack and hyperventilating, and then I tried the breathing exercise and it worked.

The three breaths doesn't take a long time. For tests in school or in sports competitions, you don't have a lot of time to sit there and meditate, and doing breathing is just a good way to calm down and put things back in perspective.

Can you imagine making this practice—three breaths, three times a day—part of your daily routine? (Circle one.)

Yes   No   Still thinking about it

We're going to end this chapter with a comforting practice that is another favorite of teens. It's called the Compassionate Body Scan, and it's really an extension of the Soles of the Feet exercise. One difference is that this is a "formal" practice, so you need to set aside time to do it—at least ten to fifteen minutes or so. It has elements of both mindfulness and self-compassion. Try it, and see what you think!

But first, let's see how you're feeling. Make an X on how your mood is, in general, right now.

Awful.
Stressed.
Terrible.

Super great!
Loving life today!

Want to say a few more words about how you're feeling? The more specific you are, the better. That way, you'll be able to compare how you feel now to how you feel after the body scan.

_____

_____

try this: Compassionate Body Scan

If you prefer to listen to the audio of this practice, you can download it at http://www.newharbinger.com/39843. Remember to take your time doing this practice. It should take you about fifteen minutes, but you can take a lot longer if you want to stretch it out.

1. To begin, lie on your back on the floor (you can do this on a bed, too, but it's a little easier on the floor). Gently rest your arms by your sides, about six inches away from your body, and allow your legs to rest on the floor. Notice your whole body as you lie there, allowing the floor to hold and support you.

2. Now, turn your attention to your breath and see if you can simply notice your breath moving gently in and out of your body. Notice how the chest rises with each in-breath and falls with each out-breath. Notice how the area around your diaphragm, below your rib cage, moves with each breath. Allow yourself to sink a little more into the floor with each out-breath. If you'd like, you can place a hand on your heart to remind yourself to bring kindness to yourself while you do this practice. Feeling the warmth of your hand on your chest, take a few deep, relaxing breaths. Then you can place your arm by your side again or keep your hand on your chest.

3. Now, shift your attention to the soles of your feet. Notice any sensations you might be feeling in the soles of your feet. Are they warm or cool, dry or moist? Take some time to really notice anything that you might be feeling there.

4. And now, shift your attention to the tops of your feet, noticing any sensations that may be there in your feet right now—maybe a tickling, or maybe the feeling of the texture of socks on your feet, or maybe nothing at all—and let every sensation be just as it is. If you notice any sensations of discomfort or even pain, see if you can allow yourself to explore the experience like a scientist observing—and perhaps mentally softening the area, imagining that you were placing a warm towel on it. And in your mind, watch it relax just a little bit.

5. Now let's have some *gratitude* for your feet. Think about it—your feet have such a small surface area, yet they hold up your entire body all day long. Isn't that amazing? Our feet work so hard for us and we rarely pay any attention to them. So take a moment to be thankful for your feet. If your feet feel good today, you can also have gratitude for that!

6. As you inhale, shift your attention from your feet up to your ankles, your calves, and your shins—paying attention like this first on one leg and then the other. Make sure that you really take time to notice the sensations in your ankles, calves, and shins. As you do so, also notice how each part of your body does so much for you.

7. When you notice your mind has wandered, simply return to the sensations in your body. Remember that the mind naturally wanders, so there's no need to judge it. And if you notice yourself judging any particular part of your body, put your hand on your heart again and breathe gently, remembering to be kind to yourself because it's hard to hear this judgment and criticism. Then return to the sensations in the body.

8. Now, returning to your legs, simply be aware of sensations of pressure or points of contact with the floor. What do those points of contact feel like? Is there any pressure or discomfort there? Or maybe something else? Take a moment to be aware of how your legs move, and how they take you wherever you need to go.

9. As you inhale, shift your attention again—now to your knees, thighs, and hips, noticing any sensations—and then pause. Maybe judging thoughts are coming up, and if so, noticing what is here in the way of those thoughts and being aware of how hard it is to hear that judging voice. If you'd like, put a hand on your heart to help you remember that you are worthy of kindness and that you can be kind to yourself in this very moment, or any moment that you need it.

10. Moving your attention now to your belly, and being aware of your body breathing. Feeling the chest expanding and contracting as you breathe, noticing how the breath nourishes you with each in-breath and soothes you with each out-breath. Letting yourself enjoy this nourishing and soothing, aware of how hard your stomach works all the time digesting your food.

11. And, now, shifting your attention to the chest, noticing your lungs expanding with each breath. Maybe you can feel your heart beating—this heart that has been beating only for you! This heart that has been beating from before you were born, when you were still in the womb—and hasn't stopped since. All for you! So taking a moment to be thankful for this heart—and these lungs—for all they do to keep you alive.

12. Shifting your attention now to your back and noticing the whole back resting on the floor, aware of any sensations present in this part of the body. Noticing places where your back touches the floor, and what that feels like right at those points of contact. What do you notice? What are you feeling?

13. As you inhale, shifting your attention down both arms into your wrists, hands, and fingers. As you move from one part of your body to another, return your awareness again and again to whatever sensations are right here in this moment. Tingling? Air moving across your skin? Warmth? Coolness? What do you notice on your arms, wrists, hands, and fingers right now?

14. Now, moving your attention to your neck, throat, and head. What sensations do you notice on your neck and head? Feelings of pressure maybe? Any signs of discomfort? This neck that supports your head all day long, and this throat that allows for you to talk, swallow, and breathe. The head and skull that contain your incredible brain, your eyes that see, your nose that breathes, your mouth that speaks, your ears that hear. So simply noticing this amazing work that goes on in your body all the time, often without your even noticing it—all the different parts that work together, in sync, that keep you alive. Pretty cool.

15. So taking a moment to put your hand on your heart again if that works for you and give yourself some love for just being here, for all the work your body does, and just because you're you and you deserve it.

16. Then gently open your eyes, slowly stretch your body, turn to your right, and sit up.

Let's see how you feel now.

Awful.	Super great!
Stressed.	Loving life today!
Terrible.	

Say a little more about how you're feeling now and what doing the Compassionate Body Scan was like for you:

_____

_____

What Teens Say

It felt like a power nap. How long was it, fifteen minutes? It felt like hours!

In the body scan, we make a point of noticing sensations in each body part—so it's a mindfulness practice. But it's also a self-compassion practice, because we approach each body part with the care that it deserves, and have some gratitude for all that each part does for us!

A few warnings about the body scan: It's very easy to drift off and fall asleep! Not that anything is wrong with falling asleep—it's one of my favorite things. But if you have homework to do, or somewhere to go, you might want to choose another meditation. Or maybe set an alarm.

my thoughts

What is your response to this chapter? Here are some ideas to think about:

- How is it that my body responds to things that I think are a threat even if they aren't a threat?

- I didn't realize that my mind wandered to keep me safe—and sometimes dwells on negative things because it's trying to help me prepare for things that could hurt me. What a thought!

- My thoughts are *not* facts?! That's shocking!

- Which activities and meditations from this chapter do I think I can realistically do in my life?

- How can I make that happen?

Write or draw below your response or reactions to this chapter.

conclusion

Mindfulness is all about seeing our experience with curiosity. It's about letting go of the thoughts and stories that run around in our heads, because often those stories are generated by our fears. Totally not our fault, but this is how we're wired. We're wired for survival, not happiness. So we have to make an extra effort to let go of the fear-based stories and see things clearly so that we can be compassionate toward ourselves. And you know what? We're not alone in this. This happens to all of us, because we're all human. Coming up in the next chapter are some ways that help us see that all of us are in this together, that we all struggle with fears, doubts, insecurities, and feeling subpar. We all have moments of feeling like a loser, and we all have moments of joy and excitement also. So if you can relate at all to feeling this way, turn the page!

chapter 6

I am not alone—even when I think I am!

You're walking down the hallway at school. Everyone you see seems to be joking, laughing, and having fun with each other. They all seem to have lots of friends—they probably get good grades, too, and if they don't, they probably don't worry about it too much. They have girlfriends, boyfriends, get invited to parties—and then there's you. You're thinking, *I've got a giant zit right at the end of my nose. I'm having a super-rotten bad hair day, and my best friend isn't speaking to me, and I have no idea why. And at the lunch table, our friends seemed to be speaking to her, and not to me. Maybe it's my imagination, but I don't think so. They seemed to be saying something under their breath to each other. It just feels weird… To top it off, I thought my braces were coming off at my orthodontist appointment this morning, and I found out I have another three months with these awful things on. So I will* have them on for picture day, *which means I can't smile for the photo, and then I'll look miserable. But that's okay because I* will *be miserable!*

Sound familiar? If so, you're not alone! Trust me, I'm writing about this here because feeling this way is familiar to most teens. So think of all the teens out there reading this book saying to themselves, *She's writing about me! This is my experience!* See? There are tons of teens out there who feel the same way you do. Whether it's a zit at the end of your nose or not getting into the college of your dreams, these feelings

of disappointment, hurt, insecurity, and sadness are shared by hordes of other teens out there who are going through just what you are. So this is what we mean by *common humanity*, which is what this chapter is about—that all teens have similar kinds of feelings and go through the same kinds of experiences. And adults do, too, actually. We're all in this "lifeboat" together.

What Teens Say

I feel like self-compassion really opened up that everyone really does have like emotional problems and everything, and it's good to feel even if other people don't talk about it or act like they don't have it, you know they do. It's comforting, I guess…

Like whatever you're feeling, you're not alone in it. Somebody else will feel the same way, will know where you're coming from, even if you think that no one understands, there will be somebody who does.

So what can we do when we feel this way—when we feel like we're all alone and we're the *only* one who is a total idiot, stupid, ugly, a failure, or the scum of the earth?

First, we can remember:

- Feeling this way is a normal part of being a teen. All teens feel this way at one time or another, and most teens feel this way a lot of the time. In fact, adults sometimes feel this way also—it's part of being human.

- Just because we think it, doesn't make it true. Our thoughts are not *facts*. They are just thoughts.

And, most importantly,

- There are some exercises we can do to remind ourselves that these difficult feelings and experiences we have are part of being human.

Here's one. It's called A Person Just Like Me. This activity, originally created by Chade-Meng Tan—though the version I use is from my colleague Trish Broderick's book *Learning to BREATHE*—allows us to get a glimpse that others are very much like us, which, ironically, we're not always aware of. In this short guided meditation, you'll first do this by thinking of a random person in your class (a neutral person), then a popular person, then someone who annoys you.

try this: A Person Just Like Me

Like other meditations, this is best done with the audio file, which you can download at http://www.newharbinger.com/39843. If you don't use the audio file, make sure that you take your time and read slowly. Allow yourself to really think about what each statement says. Remember, there's no award if you get through this quickly.

STEP ONE

Think of someone in one of your classes. She doesn't have to be a friend—it can be some random person. In fact, it's best if it's some random person.

Check the words or phrases that describe how you feel about this person:

Don't think much about her	Pretty cool
Kind of annoying	Super annoying
Like her a lot	Ugh!
Never really noticed her before	Very cool
Okay	Something else: _____

1. Get into a comfortable, relaxed position.

2. Try to get a good image of this person in your mind. What does she look like? What's the expression on her face? What is she wearing?

3. Let's consider a few things about this person:

   This person is a human being, just like me.

   This person has a body and a mind, just like me.

   This person has feelings, emotions, and thoughts, just like me.

   This person has, at some point, been sad, disappointed, angry, hurt, or confused, just like me.

   This person wishes to be free from pain and unhappiness, just like me.

   This person wishes to be safe, healthy, and loved, just like me.

   This person wishes to be happy, just like me.

4. Now let's allow some wishes for this person to arise:

    I wish for this person to have the strength, resources, and support to help her through the difficult times in life.

    I wish for this person to be free from pain and suffering.

    I wish for this person to be strong and balanced.

    I wish for this person to be happy,
    because this person is a fellow human being, just like me.

5. Take a few more deep breaths and notice your experience. Take at least a couple of minutes to just sit here and think about how you feel.

Now what do you notice about how you feel about this person? Circle any words or phrases that describe how you feel about them. Be honest—no one is checking!

Don't think much about them   Pretty cool

Kind of annoying   Super annoying

Like them a lot   Ugh!

Never really noticed them before   Very cool

Okay   Something else: _____

If something changed about how you felt about this person, why do you think it changed?

_____

_____

_____

STEP TWO

Do this same practice, but this time think of the most popular kid in school. The one that you think everything seems to just go well for, who seems to have lots of friends, and just seems to roll with everything.

Get a good image of him in your mind…the beautiful hair…the flawless skin…the friends gathered around.

And do the same practice, A Person Just Like Me, with this person in mind.

1. Get into a comfortable, relaxed position.

2. Try to get a good image of this person in your mind. What does he look like? What's the expression on his face? What is he wearing?

3. Let's consider a few things about this person:

    This person is a human being, just like me.

    This person has a body and a mind, just like me.

    This person has feelings, emotions, and thoughts, just like me.

    This person has, at some point, been sad, disappointed, angry, hurt, or confused, just like me.

    This person wishes to be free from pain and unhappiness, just like me.

    This person wishes to be safe, healthy, and loved, just like me.

    This person wishes to be happy, just like me.

4. Now let's allow some wishes for this person to arise:

    I wish for this person to have the strength, resources, and support to help him through the difficult times in life.

    I wish for this person to be free from pain and suffering.

    I wish for this person to be strong and balanced.

    I wish for this person to be happy,
        because this person is a fellow human being, just like me.

5. Take a few more deep breaths and notice your experience. Take at least a couple of minutes to just sit here and think about how you feel about this person now.

What was this experience like? Has anything changed about the way you think or feel about this person?

_____

_____

_____

If something has changed about the way you feel or think about him, why do you think it has changed?

_____

_____

_____

STEP THREE, IF YOU DARE

Now, try this for the advanced practice—only for the hard core out there. You definitely don't *have* to try this one!

Do this same practice, but this time think of someone who *really* annoys you. This doesn't have to be someone you really dislike, but it could be—it could also just be someone who bugs the crapola out of you.

Get a good image of her (or him) in your mind. (I know it's tough—hang in there!)

And do the same practice, A Person Just Like Me, with *that* person in mind. Here are the words one more time. Remember not to rush through this practice, although it might be tempting!

1. Get into a comfortable, relaxed position.

2. Try to get a good image of this person in your mind. What does she look like? What's the expression on her face? What is she wearing?

3. Let's consider a few things about this person:

   This person is a human being, just like me.

   This person has a body and a mind, just like me.

   This person has feelings, emotions, and thoughts, just like me.

   This person has, at some point, been sad, disappointed, angry, hurt, or confused, just like me.

This person wishes to be free from pain and unhappiness, just like me.

This person wishes to be safe, healthy, and loved, just like me.

This person wishes to be happy, just like me.

4. Now let's allow some wishes for this person to arise:

I wish for this person to have the strength, resources, and support to help her through the difficult times in life.

I wish for this person to be free from pain and suffering.

I wish for this person to be strong and balanced.

I wish for this person to be happy,
    because this person is a fellow human being, just like me.

How do you feel about this person now? Has anything changed? Remember, be honest!

_____

_____

_____

If you feel differently toward her, why do you think you do?

_____

_____

_____

So what have we learned from this guided meditation? When we look beyond our own normal perceptions, when we take down the wall that separates us from others, we see that others—whether they're the random kid in your class, the popular kid, or the kid that you have a hard time with—are really not so different from us. We all have similar fears, insecurities, doubts, and desires to be happy and liked by others. We all struggle at times. So we're all here trying to surf the waves and trying to find ways to keep afloat, and maybe even ride a wave or two.

So the next time you encounter that annoying person, you might want to quickly do the A Person Just Like Me meditation. It really helps!

Personal Story from a Teacher

Dr. Mark Greenberg, a professor and researcher who studies youth, parents, and the emotional kind of learning that happens in schools, says, "I use A Person Just Like Me all the time at the beginning of meetings with adults. I often find that adults seem to forget that others feel the way they do, that others also struggle and experience the same kinds of challenges that they do. It helps to remind all of us that we're all in the same boat. And then the meetings seem to go a little bit better."

Now maybe you're thinking, *But don't my "flaws" make me different and deficient and set me apart from everyone?*

Au contraire—that's French for something like "No, it's quite the opposite." We all have "flaws." It's part of being human. It actually is something that connects us with others, because it's something we have in common. But "flaws," which you notice that I've put in quotes (because I don't necessarily consider them flaws— and you may not either by the time you get to the end of this chapter), are not necessarily such a bad thing—they don't make us deficient. *Au contraire*, they make us delightfully special and unique.

So before we go on, think of what you might consider to be your flaws. My guess is that you probably don't need too much time to think of them.

Now draw an X on this line to indicate how bad you think your flaws are:

Really awful.   Not so bad!

In our culture, there's a lot of emphasis on doing things "perfectly" and being perfect, or at least not seeing our flaws as so bad—in other words, being as far on the right side of this line as possible. But not all cultures are like that. For example, there's a very interesting kind of Japanese art called *kintsugi*. When pottery breaks, rather than just putting it back together to make it "as good as new" by hiding the cracks, *kintsugi* artists fill the cracks with gold, making the object stronger and "better than new." The gold actually makes the cracks stand out and "shine." And the message we can get from this is that, like pottery bowls, we can also honor and embrace our own "imperfections" and "flaws." And also, as we are all imperfect beings, it connects us to one another.

try this: Make Your Own Kintsugi Bowl

Materials you will need:

A disposable paper bowl that is *not* wax covered. Eco-friendly bowls work best. (You can find them at www.amazon.com.)

A black pen that writes on the bowl. A gel pen works well.

A gold permanent marker or paint pen (any store that sells art and crafts supplies has them)

First, draw cracks on your bowl with the black pen, so that it makes sections. Your bowl should look something like this:

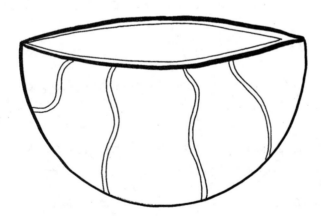

Now you can fill each section with a different "mindful design." Remember that when you do this, really take your time so that you can focus—think about the feeling of the pen in your hand, the feeling of the pen as it touches the bowl, watching as the ink comes out of the pen onto the bowl, noticing the spaces in between the black lines of your design, and so forth. The goal is not to get done with the project, but to *notice* and *observe* sensations. Here is an example of what it might look like—but, of course, don't worry if yours doesn't look like this. Yours is supposed to look like *yours*, not ours:

Once you're finished with drawing designs in the sections of your bowl (or if you get tired and don't want to do any more drawing), use the gold marker to draw over the black lines. Voilà! Your *kintsugi* bowl is complete, and you can use if for whatever you'd like. Whenever you see it, however, keep in mind that our imperfections, our uniqueness, is what makes us special and, at the same time, connects us with all other humans on this planet. And when we make mistakes, which we all will do over and over throughout our lives, we can use those mistakes to learn and make us stronger.

What Teens Say

I liked the [**kintsugi** bowl] because it symbolizes the goodness that can be found in mistakes.

I like the idea that the gold in the cracks makes the bowl stronger. So the more cracks you have, the stronger you are!

Now draw an X on this line indicating how you feel about your flaws:

Really awful.   Not so bad!

my thoughts

Take a moment to think about anything you may have learned from this chapter. Feel free to draw or write about your reflections here. For example, you might want to think about what it's like to see that others also struggle and deal with feeling lost, hurt, angry, or alone.

conclusion

Often when we're struggling, we feel very much alone. We feel deficient, defective, and inadequate. And because we get the message from our culture that we're supposed to be "perfect" and it's not okay to have "flaws," and because we're taught from an early age that there's something wrong with us if we feel sad or angry, we learn that we should hide these feelings and these "flaws" from others. The truth is that struggling and feeling this way happens to all of us, especially when we're teens. And these "imperfections" and "flaws" that we all have—they're part of being human, part of being a living, breathing human being. They don't make us defective or inadequate in a way no one else is—if you think about it, it's actually something that connects us to everyone else in the world because it's something that we all share.

Part 3

Beyond the Basics

chapter 7

help! I'm on an emotional roller coaster!

I will never, ever be good enough for them! Never! No matter what I do, I'm still a loser to my parents. So what if I fought with my bratty sister? I mean, really, she started it. Okay, so what if I may have teased her a little? I know I shouldn't have, but here I am at the end of the worst day of my life, and then they make me feel like I'm the lowest of the low. I mean, I missed that soccer goal and lost the game for my team, I spilled the stupid acid in chemistry lab, squirted ketchup all over my shirt at lunch, and to top it off, they make me feel like I'm the worst kid in the world . . . I mean, I already know I am . . . I know I royally messed up today and I'm a total idiot for it all. I'm sorry I'm a loser! I hate myself! I know I'll never amount to anything! I know I'm the worst of the worst! Okay? Does that make you happy now?!

Sound familiar? As you know by now, you are not alone, and this is what many teens go through with their parents as roles and relationships shift during the teen years. You've learned some ways to practice self-compassion—by setting aside time to do practices each day. But what happens in the midst of your day when you're minding your own business and something happens that suddenly jolts you, and that old nasty habit of beating yourself up creeps in? In this chapter, we're going to learn a practice that is particularly helpful when that emotional roller coaster really takes us for a ride and we've taken a dive into that dark tunnel of self-criticism.

Remember in chapter 1 we talked about how the teen brain goes through massive changes—pruning away the neurons and neural pathways that are not being used and creating new neural pathways? Also—remember that this is pretty key when it comes to feeling overcome by emotions—your amygdala, that part of the brain responsible for feeling fear and anger, becomes developed more quickly during teen years than the prefrontal cortex, the part of your brain responsible for thinking logically and making careful decisions. So, of course, you feel huge waves of emotion from time to time! Your amygdala is like a superpower compared to your prefrontal cortex!

Dr. Laurence Steinberg, a researcher who has worked with teens for decades, often says that it's like having a car with an accelerator and no brake!

So when you're feeling these huge waves of emotion, when you feel like there's this mammoth tidal wave coming at you and it's about to pull you under, remember it's not your fault. It's just your teen brain doing the best job it can and responding to the situation at hand.

And one day, the prefrontal cortex part of your brain is going to catch up with the amygdala, and things will level out a bit more. And your tidal wave will go from this:

To this:

There still will be waves, but they will be much smaller and more manageable.

Have you ever felt like you were being swamped by a tidal wave of emotion? (Circle one.)

      Yes   No

How did you respond? (Circle any of these.)

Ate too little	Got violent
Ate too much	Isolated myself
Binged out on TV series	Refused to go to school
Cut myself	Said things that weren't nice to people
Did drugs	Screamed a lot
Drank too much	Sobbed
Got really, really depressed	Stormed out of the room

As you've probably realized, these responses are not all that helpful in the long run, although they might have helped you feel better for a little while.

So what else can you do? While the tidal waves are looming, what can you do to stay afloat and not go under?

The meditation practice I'm going to teach you in this chapter—the Soften, Soothe, Allow practice—will help in these situations. Before you do it, though, let's try this art activity. You don't absolutely have to do this art activity before you do the meditation practice, but it helps because it gives you an actual "feel" for what you can do when you feel strong emotions in your body. You'll soon see what I mean.

Warning: This art activity can get pretty messy, so please make sure you clean up after yourself!

try this: Playing with Oobleck

You might be familiar with Oobleck from your preschool days!

You'll need:

1 cup water

About 2 cups cornstarch (must say "cornstarch," not cornmeal, cornmeal mix, corn flour, corn dogs, or any other kind of corn)

Mixing bowl

Wooden spoon for mixing

Plastic container with a lid,

1. Mix 1 cup water with roughly 2 cups cornstarch in a mixing bowl. To reduce cleanup later, it's best to do this over the kitchen sink or outside.

2. Keep mixing with a wooden spoon for about five or ten minutes until you get it to a consistency where you can take a piece of it and make a semifirm ball, but when you open your hand it "melts" into a flat pancake.

3. When you get close to the right consistency, you'll likely need a bit more water. Add a few drops at a time, mix, and then if you need a little more water, add a few more drops and mix. If it feels too loose (you can't get it to form a semifirm ball), add just a teeny bit of cornstarch at a time—maybe around a teaspoon.

4. Now play with the Oobleck. Take a fistful and hold it tightly. Notice what happens. How would you describe this? (You might want to wait and write about this later when your hands are clean!)

_____

_____

5. Open your fist and hold your hand out, palm facing upward. Describe what happens to the Oobleck in your hand:

_____

_____

6. Do this a couple of more times—tightening your fist around the Oobleck (it should form a ball that feels pretty solid) and then opening your fist with your palm facing upward (the Oobleck should "relax" and kind of melt and flatten in your hand).

7. When you're done, store your Oobleck in a closed plastic container so that it doesn't dry out too much, in case you want to play with it later.

Now that you've played with the Oobleck and you know what it feels like, let's try that exercise I was talking about that will help you work with difficult emotions. The Soften, Soothe, Allow practice can be done slowly like a meditation, and then once you know the different parts and how to do it, you can do it "in the moment" when you're feeling the tidal wave of emotion coming on. But for now, it's important to really take your time and do it slowly, giving yourself lots of time to notice the different feelings that arise.

try this: Soften, Soothe, Allow

If you would like to listen to instead of read the exercise, please download the audio at http://www.newharbinger.com/39843.

Find a comfortable position, close your eyes, and slowly take three relaxing breaths. Place your hand on your heart or give yourself another soothing touch gesture (which you learned in chapter 3) to help you remember to bring kindness to yourself.

Now think about a mild to moderately difficult situation that you are experiencing in your life right now. This might be a problem you are having with a friend or a parent, or maybe a problem at school. Are you feeling bullied? Unappreciated? Left out? It's best at this point not to choose either the most difficult problem you have or the tiniest problem. You want to choose a big enough problem so that you can feel a little stress in your body when you think of it, but not something too huge, because you're just getting started with this practice.

Clearly visualize this situation. Who was there? What was said? What happened? Try to get a good image of this situation in your mind.

Write about the situation here if you'd like:

_____

_____

_____

_____

How full is your "emotional cup" when you think of this situation? Shade in how much emotion you are feeling right now when you think of this situation. How high up in the cup does your emotion rise?

Now that you're thinking about this situation and the emotions it brings up for you, see if you can label the feelings you have about it:

1. Name each of the different emotions or feelings that fill your emotional cup. Sometimes it takes some time to open up to and feel these emotions. So remember not to rush. Circle any of the feelings that you are experiencing when you think of this problem:

    Anger

    Confusion

    Fear

    Frustration

    Grief

    Hurt

    Loneliness

    Sadness

    Something else? (Write it here): _____

2. Now, see if you can name the difficult emotion that's strongest. Is there one feeling that rises to the surface? Write that emotion below. (And if there's a tie between two feelings, just choose one.)

    _____

3. Repeat the name of the emotion to yourself in a gentle, understanding voice, as if you are sympathizing with what a friend of yours is feeling: *That's hurt. That's sadness. That's embarrassment.*

4. Use the same warmhearted tone of voice that you would use with a friend. If you said to your friend, "Wow, you're really feeling bad," what tone of voice would you use? How would you say it? Use this same kind voice with yourself right now.

The next steps of the practice involve what I call "feeling it in the body."

5. Now expand your awareness to your body as a whole.

6. Recall the difficult situation again and scan your body for where you feel it the most. In your mind's eye, sweep your body from head to toe, stopping where you sense a little tension or discomfort. Take your time doing this. You're simply investigating all parts of your body, noticing if there's a place where there's tightness. Sometimes this feeling is

really noticeable, but sometimes it's subtle. Where in your body do you feel this emotion? If you can feel it in a few places, you can write them here:

_____

7. Now choose just one place in your body where the feeling seems to be the strongest. Perhaps you notice a tense muscle or an achy feeling—like a heartache, or a burning or gripping in your stomach, chest, or even throat. Sometimes you feel it in your head. Notice how that feels right now. Take time to allow yourself to feel the discomfort. Imagine yourself turning toward the discomfort rather than pushing it away, which is what we often do. Count slowly to ten while you allow yourself to notice this feeling. If you want, you can stay with the feeling longer. Just notice how it feels in your body. Describe what this feeling is like:

_____

_____

_____

And here's the final step: "soften, soothe, allow."

8. Now, soften around the area where you feel the sensation in your body. Imagine you're taking a warm washcloth and putting it on that area where there is discomfort, letting the muscles soften in this area. Remember the way the Oobleck softened in your hand when you opened your fist? In the same way, you're "loosening your grip" on this discomfort in your body—you're relaxing around it a little. You can say quietly to yourself, *soft . . . soft,* and imagine it softening just the way the Oobleck let go. Remember that you are not trying to make the sensation go away—you are just holding it softly, like you might hold a baby chick—very delicately and with a lot of tenderness.

9. If you just want to notice that the emotion is there and you don't want to soften at all, that's fine. Just naming the emotion can be helpful. You can take your time working with the sensations in the body.

10. And if at any point the uncomfortable feeling gets to be too much, you can always go back to just noticing your breath as you breathe in and out, simply staying with that.

11. Now, saying some soothing words to yourself just because you've been struggling in this way with this strong emotion. It's hard, isn't it, to feel these really difficult feelings? Put your hand over your heart and feel your body breathe. You might want to actually whisper kind words to yourself, such as, *This is really hard* or *I know I'm not the only person who has*

ever felt this way. What kind words work for you? What words would help you feel a whole lot better right now? If you'd like, you can write these words or phrases below:

_____

_____

_____

12. Just allowing the strong feelings and sensations to be here. Making room for them. Maybe even letting go of the wish for this uncomfortable feeling to go away. Remembering what it was like when you opened your hand when the Oobleck was in it. The Oobleck sort of relaxed and melted a little bit, didn't it? This is what we're doing—just opening and allowing whatever feelings are here to just be here, without any expectation or wish for them to go away. And maybe watching them loosen up a bit.

13. These three words—"soften…soothe…allow," "soften…soothe…allow"—can be repeated, kind of like a favorite slogan or mantra, reminding you to offer yourself a little kindness and warmth in these moments of difficulty.

14. If you ever feel too much discomfort with an emotion, you can always simply stay with your breath until you feel better.

Now we're going to see how this practice was for you.

What happened when you "labeled" your feelings—when you actually whispered silently to yourself, *That's anger* or *there's hurt*? (Circle one.)

I felt a little less of the emotion (less angry, less hurt).

I felt a lot less of the emotion—it almost seemed to go away.

Nothing changed.

Something else: _____

All of these responses are totally normal. Our feelings are simply that—our feelings. We're not supposed to feel one way or another.

When we label our emotions, we engage our prefrontal cortex, the part of the brain that's responsible for logical thinking. When this happens, it often quiets down our amygdala, the part of the brain that becomes activated when we're feeling strong emotions. So you may have noticed a little shift when you labeled your emotions—maybe the emotions became a little lighter or a little less powerful. "Name it and you tame it" is one way to think about what happens when you label emotions.

Were you able to find where in the body the emotion was? (Circle any places that you felt your emotion might be.)

Back of neck

Chest/heart area

Face

Forehead

Head

Shoulders

Stomach

Throat

Someplace else: _____

What happened when you tried to soften and soothe the area where there was discomfort? Did you notice any slight shifts or changes in how you felt? What was this like?

_____

_____

_____

How about when you "allowed" the feeling or sensation to be there without trying to make it go away? What did you notice?

_____

_____

_____

Now here's your emotion cup again. How full is it? Shade or draw how far up in the cup your emotion goes.

Was there any difference in the emotion cup before and after you did this practice? Even slight differences count. Even if it is just an eighth of a cup difference, it's that much less that you feel upset or sad.

What Teens Say

The naming your emotions [practice] is very helpful for me because I generally get very overdone with emotions.

I'm very empathetic, so reading books, walking on the street, listening to depressing songs, I get so depressed, so just having that...

The next time you feel overwhelmed with emotion, try the Soften, Soothe, Allow process the way these teens did. Do it right in the moment when you are upset, like a kind of quick fix-it tool: name the emotions, find them in your body, and then soften around them, soothe yourself, and allow the emotions to be there. Remember, we're not trying to make the difficult feelings go away—we're just allowing them to be here and making room for them, and then seeing what happens.

After you've tried the exercise, come back here and write about the experience in the space below. Were you able to make space for the uncomfortable feelings and allow them to be there?

_____

_____

_____

Remember, anytime you do this practice, you're just seeing what happens, and whatever happens is okay.

my thoughts

Do you have any responses or reactions to this chapter?

Use the space below to write or draw about any thoughts you have:

conclusion

Feeling overwhelmed by feelings, particularly feelings like anger, hurt, or loneliness, is part of being a teen. It's something that most, if not all, teens experience from time to time, and a lot of teens beat themselves up for it. But that doesn't mean that you have to drown in the tidal wave of emotions. There are specific practices that you can learn (and keep with you in your back pocket) that will allow you to bob to the surface—and learn to surf.

Like overwhelming emotions, another thing that can be really stressful and frustrating is when we feel like no one is listening to us, that no one really hears us. In the next chapter, I'll show you things that you can actually do to get others to really, *really* hear you.

chapter 8

communicating with mindfulness and compassion

The most frustrating thing for so many of us is feeling like we're not being heard. Or that we're not being understood. Or that we're being ignored. Or not seen for who we are. It may feel to you like your parents treat you like they did when you were ten, and that's not who you are now! You've changed! And teachers have their own agenda, and can rattle on without really listening to your thoughts or opinions about things.

So what can we do to make sure that people hear us—and I mean *really* hear us?

First of all, although we don't have ultimate control over whether or not we're really heard—we never have control over someone else as much as we'd like to— when we speak and listen openly from our hearts, which is what we do when we speak with compassion, it's more likely that people will hear us.

Let me give you an example. Read below:

Friend #1: You should never have told Mario that I liked him! That was a secret! And it was my business! You're such a big mouth! I told you not to say anything! Whatever possessed you do that?

How would you feel hearing this from a friend?

_____

_____

What would your response to this be?

_____

_____

How does your response make you feel?

_____

_____

Friend #2: I know that you didn't mean any harm, but I would have preferred if you asked me first if it was okay to tell Mario that I liked him. I'm feeling a little embarrassed now.

Notice how you're feeling this time. How would you describe this feeling?

_____

_____

What would your response to this friend be?

_____

_____

How does your response make you feel this time?

_____

_____

When someone comes at us in attack mode, as a means of protecting ourselves, we're going to shut down and not listen. We're going to put up our guard. But if someone comes to us in a much more gentle and compassionate manner, we're going to be a lot more likely to pay attention and hear him. The same thing happens when we approach someone else. If we are really, really listening and open to what she has to say—in other words, we're compassionate—it's much more likely that she will really hear us and see us, and that she'll be compassionate also.

So how do we make sure that we approach others in a gentle and compassionate manner and not be in a reactive attack mode with our fists up and ready to fight?

The next two practices will help. I'm going to offer a series of steps (inspired by the process outlined in *Insight Dialogue* by Gregory Kramer) that are helpful in getting you to a place where you can communicate clearly, truthfully, openly, honestly, and mindfully.

try this: Listening to My Heart

If you prefer to listen to this exercise rather that read it, please download the audio at http://www.newharbinger.com/39843.

1. Find a comfortable place to sit or lie down, a place where you feel safe and relaxed.

2. Think of something that is pretty important to you that you would like to communicate to someone else, perhaps a parent, teacher, or friend. Maybe you haven't yet talked to this person about this because you have been too scared or shy, or you've been too angry or hurt to talk about it. Take some time to think about this.

3. If you'd like, you can write what you'd like to say here:

    _____

    _____

    _____

4. Now, stop for a moment. Notice your body as you are sitting or lying here. Get an image of your body as you are sitting or lying down in this room. Notice where your body touches the chair or couch, and bring your attention to what you feel at that point of contact—are you tight there or stiff? Do you feel pressure or even discomfort? Do some areas feel different than others? See if you can view your body as a whole as you are here, in this moment. Get a mental picture of your whole body. As always, take some time, at least a couple of minutes, to do this.

5. Next, relax, and rest in the moment. Take a few minutes to scan your body from head to toe and see if there are any places where you feel tension or tightness, or even pain. If you do, see if you can relax and soften those areas. Are your shoulders stiff? Allow them to relax. Is your forehead furrowed? Let those face muscles go. How about your jaw? Is your jaw relaxed? Is your tongue relaxed in your mouth? Taking your time, resting here.

6. Now, when you find some tense place in your body, can you turn toward it with acceptance? This is sort of like imagining there's a door to your uncomfortable area—say, in your stomach—and you are opening that door and allowing your stomach to just be there. Can you do that with other parts of your body that might be a bit tense? Just allowing all parts of the body to be there, not pushing them away. Bringing a bit of comfort to those uncomfortable areas as much as you are able to.

7. Now turn your attention to the external world. Notice any sounds around you—sounds that might be near or far away—soft sounds, louder sounds. What do you hear? Spending some time, just listening—no rush. There's no place to go, and nothing to do other than to listen.

8. Finally, listen to your heart. There is something that you'd like to tell someone; your heart will know what that something is. Write the words of your heart in the space below. Take all the time that you need.

    _____

    _____

    _____

_____

_____

_____

_____

_____

_____

_____

_____

9. Now imagine that the other person is _really_ listening to each and every word and really hears you. What would he say in response? Put yourself in his shoes. For example, if he is a parent, think of what it might be like to be your parent, to want to make sure you're safe, but also to want you to be happy. If it's a friend, think of what it might be like for him, to also be trying to make it through these years, feeling insecure and unsure a lot of the time, and sometimes stumbling, sometimes making mistakes.

Listen with your whole attention to his response, to what he has to say. Remember that he, too, is speaking from his heart, in the most honest and open way that he can. What does he say?

_____

_____

_____

_____

_____

_____

_____

_____

10. Now if you'd like, you can respond to what he's said. Remember that you are listening deeply, and really hear him. Is there anything that you'd like to say back to him? If so, you can respond here.

_____

_____

_____

_____

_____

_____

_____

_____

_____

_____

_____

11. Now letting go of the conversation, coming back to feeling your body as you are—here in this space, in this room, in this moment. Notice the contact of your body with the chair or couch. Notice what that feeling is like, and whether it is the same all over, at all the points of contact. Keep noticing those points and see if they change at all. Continue in this way for several minutes.

What stood out to you most in this exercise?

_____

_____

_____

_____

Were you surprised by anything that came from your heart? If so, what was it?

_____

_____

_____

Were you surprised by anything that came from the other person's heart?

_____

_____

_____

You can see that listening is a huge part of mindful communication. And not just listening the way we normally listen—which is really one part of us only listening a little, while the other part is thinking about what we're going to say next. Really listening is being present with our full attention—listening deeply, hearing every word and even the inflection and intention behind every word.

What Teens Say

If we really paid attention to what others were saying, we would be closer. If you aren't really listening to someone, you can't understand what they're saying. If neither of you is paying attention to the other, it's almost like two separate conversations going on.

A good way to practice listening deeply that is also super enjoyable is by listening to words, or lyrics, in music. Remember the Music Meditation exercise that we did in chapter 5? This is a somewhat different way of listening to music. Check it out at http://www.newharbinger.com/39843.

try this: Speaking So Others Will Listen

The secret to speaking so that others will really hear you is to speak from a nonreactive place, no matter what you're feeling. Whether you're angry or upset at something someone just said, or terrified to speak up, or just want to express your opinion about something that's important to you, the solution is the same.

So how do you do this?

Here's the key: Make sure that you're centered and calm before speaking. And you do this through practicing mindfulness and self-compassion in the moment.

Say someone just said something that angered or upset you. For example, your parents won't let you go to a party that you've been looking forward to because you didn't mow the lawn.

Exercise self-compassion by doing the practice A Moment for Me: engage in a soothing touch gesture like putting your hand on your heart. Now say to yourself:

> *This is a moment of struggle. This is hard! I really wanted to go to that party! I'm really disappointed! And I'm super angry!*

> *Struggles are part of life. We all struggle—I am not alone. Lots of people feel angry and disappointed.*

> *May I be kind to myself in this moment* (and then say some kind words to yourself like you would say to a friend who is in this situation). Maybe *I know how hard this is for you. So disappointing. So upsetting. I'm here for you.*

Do this until you feel calm and ready to communicate in a clear way—without arguing and pleading, because that won't help them listen.

What if you're terrified of speaking up? Let's say someone has been doing the "mean girl" thing with you at the lunch table, ignoring you and whispering behind your back, and you want to tell her to stop, but you're shaking in your shoes.

You can get centered and calm by doing the Soles of the Feet exercise or a three-breath practice.

For three-breath practice, take three mindful breaths. Slowly breathe in and out, noticing the feel of your breath. If your mind wanders, bring your attention back to your breath.

Continue feeling your breath in this way until you feel calm, centered, and ready to speak.

Once you start speaking, if you start to feel upset or you notice that your voice is rising, come back to the feeling of a physical sensation—your feet on the floor, your butt in the chair, your hands clasped, or your breath. Always come back to physical sensations.

After the conversation is over, if it's been a bit difficult—and especially if you find yourself being self-critical about how it went—you can always do one of your favorite self-compassion meditations. Even a simple hand on your heart can remind you that this is hard, and you deserve kindness.

Same thing when you have something that's really important to you and you want to express your opinion. Maybe you're getting into a heated debate in history class, and you really want others to hear you. Choose a practice that will help you be less emotional so people will be more likely to pay attention to what you have to say. After a while, you'll get the hang of which practices work for you in which situations.

What Teens Say

I'm able to communicate with people I'm scared of more.

try this: Keeping Track of Challenging Conversations

On the next page is a chart that will help you keep track of when conversations go smoothly and when they don't. You can keep track of what happens differently each time. And remember when things don't go well that you are human, that you are allowed to make mistakes. Be kind and gentle with yourself, and do a self-compassion practice!

Who was this conversation with?	What happened?	How did I feel after it was over?	How did I treat myself after it was over?	If it did not go well, what self-compassion practice will I do right now? How about the Music Meditation exercise? Or simply try the practices Soothing Touch or A Moment for Me from chapter 3?

One important point to remember is that we don't have ultimate control over what other people say in conversations. All we can do is do our best and "set the stage" so that the other person is more likely to listen to us. After that, we have to let go of what happens—and be kind to ourselves if the conversation doesn't go the way we'd like, which it won't always. Remember, we're human, and this is part of the way life as a human is. Sometimes it goes our way, and sometimes it doesn't. And we can be kind to ourselves throughout.

my thoughts

What are your reflections about this chapter? Here are some questions just to get you thinking:

Was it different for you to listen deeply? What was it like when you spoke from your heart? Is this different from the way your conversations with others normally are? If so, do you think it would work for you to communicate with others in this way? Why or why not?

_____

_____

_____

_____

_____

conclusion

In life in general, we spend much time in conversation, but unfortunately, we don't necessarily speak the truth from our hearts or listen deeply. So often our communication with others comes from our own anger or fear that we are not being understood, not heard, and not acknowledged or respected for who we are. When we slow down, take time to really speak from our own hearts and listen deeply to others, we can communicate in a much more honest way—and usually this ends up being a lot more effective in getting others to hear us.

So often others' responses to us affect us in a way that hurts our self-esteem. In the next chapter, you'll read about how self-compassion is different from self-esteem—how it is there for you when self-esteem crumbles. So read on!

chapter 9

what about self-esteem?

You've probably heard lots about self-esteem. But self-esteem is very different from self-compassion. Both involve an attitude we take toward ourselves, but that's pretty much where the similarity ends. We've learned what self-compassion is—it's treating ourselves with kindness when we're feeling bad. *Self-esteem* is how we understand our own worth and value in the world—it's like self-confidence. And we've all heard about how it's important to have self-esteem—how if you have it, you'll feel better about yourself, you're more likely to do well at school, and more likely to have lots of friends. If you're like how I was as a teen (about a zillion years ago), you probably have some times when you feel pretty okay about yourself, and other times when you feel like a total turkey and you can't imagine why anyone would want to be friends with you. Teens often spend a lot of time wondering how you go about getting this feeling of being valued and confident—and why others seem to have it and you don't. You may also wonder what good self-esteem would do you anyway when you're feeling rotten about yourself.

So, to understand this, it's helpful to know a little bit about the history of self-esteem. In the last generation or so, we thought having self-esteem would be the answer to all our problems. We thought that if we could raise kids' self-esteem by praising the things they did well, they would feel better about themselves, and that would help them to do really well at school, and grow up to be well-adjusted, confident, and successful adults.

Guess what? Some of that has turned out to be true, but a lot of it hasn't. When feeling good about yourself is hinged on doing well, you're treading in dangerous territory. For example, when you make a winning goal at soccer, you feel great about yourself. But what happens when you don't make that goal—when you miss? Missing a goal is just human, even for a superstar athlete. But your self-esteem, or self-confidence, does a deep dive, and you feel unworthy and like a crumb, right? So although self-esteem is generally a good thing, there are some problems with how it works.

Let's see what happens when we base how we feel about ourselves on self-esteem. This next exercise will show us what can happen.

try this: Seeing Self-Esteem for What It Is

This exercise has two parts. The first explores what happens when we base self-esteem on how we perform, and the second talks about how we actually get self-esteem.

First, think of a time when you didn't do well at something that you considered yourself pretty good at. (This happens to all of us, by the way, so don't sweat about it.) What was the situation?

_____

_____

_____

How did you feel afterward?

_____

_____

So if we base how we think and feel about ourselves by how we perform, then our sense of ourselves is *conditional*—it can rise or fall depending on external circumstances, sometimes circumstances that are totally out of our control. This sense of self is pretty unstable, right? That's a problem, because we'd like to feel good about ourselves all the time.

And here's the second part. It has to do with how we actually get self-esteem.

For a moment, think about something that you know you do pretty well. It can be something like doing well in math, or shooting hoops, or playing an instrument, or it could be something like being a good brother or friend, or maybe it's being organized or on time to meet people. Write the thing that you know you do pretty well here—and no one else is reading this, so you can be honest!

_____

Now, think about how you know that you're good at this thing.

(You might need some time to think about this—take your time. Remember, it's okay to be slow about this, because this stuff takes some thought.)

_____

_____

My guess is that in some way, you compared yourself with others, right? For example, you know you're good in math because you do better than others, and you know you're pretty well-organized because somewhere in the back of your brain you're thinking of others who aren't as organized as you. Or you know you're good at basketball because you get more baskets than most people, right? So although it feels good when we know we do well at certain things, when we compare ourselves with others in this way, it actually separates us from them in our minds, and may make us feel a bit "above" them, right?

It's the opposite of common humanity (see chapter 6). It causes us to feel isolated and more alone. Even when we feel proud of our achievement, in a subtle way, we feel more distant from others. And it's really important for us to feel connected with others—it's maybe what we need more than anything, as we learned in the last chapter about how important it is to be heard. So this is another problem with self-esteem.

Guess what? I've got some good news for you. Self-compassion offers an alternative in how we can relate to ourselves—a way that is more stable and dependable, a way that we can relate to ourselves that is not based on our achievements or whether we are successful in the moment. Self-compassion means being a good friend to yourself—and not just when things are going well but *especially* when you're having a hard time.

What Teens Say

Self-esteem says, "It's okay, you did this good," which they then connect to "I'm a good person." Self-compassion says, "It's okay you did this poorly. You messed up, but it's okay."

Remember in chapter 3 where you related the kind words that you said to your good friend when he had a rough day? And the not-so-kind words that you said to yourself? And remember that we talked about the difference?

Well, it doesn't have to be this way. We can be kind to ourselves, too, just as we are to our friends. We all have kindness within us—it comes with the territory of being human. We *can* be compassionate with ourselves; we're just out of practice. So it takes some getting used to.

We give ourselves compassion (kindness) not necessarily to make ourselves feel better—although, of course, that would be nice (and by the way, it usually happens along the way). We give ourselves compassion simply because we are feeling bad.

So self-compassion is about being with ourselves when we're feeling bad, the way a comforting parent, grandparent, or friend is with us. It's like putting on a warm blanket when we're feeling chilled, or sitting with a toasty mug of hot chocolate in front of a fire after we've been outside in the cold. It's about taking care of ourselves by figuring out what we need, and then giving ourselves exactly what we need.

And we all have the capacity to do that. We all have the ability to be kind to ourselves within us—we're just not used to doing it. There are tools to help you, so read on.

This practice is a favorite of teens because you get to use your imagination: you can just lie back, get comfortable, and imagine that you're going on a magical voyage to a special place.

But before we get started, it might be helpful to describe how you feel before you start this practice.

You can use your own words or any of these: terrible, pretty good, not bad, so-so, feeling great today, happy as can be, depressed, sad, angry, frustrated, terrific, jumping for joy, the worst ever, super, awful, so sad!! Remember that you don't have to feel any particular way—whatever you're feeling is simply what you're feeling.

_____

On a scale from 1 to 10, with 1 being feeling really awful and 10 feeling really great, what number would you use to rate your mood right now? _____

Okay, ready?

try this: Compassionate Friend

If you prefer to listen to rather than read this practice, please download the audio at http://www .newharbinger.com/39843.

1. Allowing your eyes to close, begin by taking a few deep inhalations and allow your shoulders to relax away from your ears.

2. Now, turning your attention toward your breath, just feel your body breathing—in and out.

3. Take a few moments to allow yourself to imagine a place where you feel safe, comfortable, and relaxed. This can be a real or imagined place—but a place that allows you to breathe comfortably and let go of any worry. Perhaps this place is in nature—at a beach, in the woods near a brook—or maybe it's a corner of your bedroom or the comfort of a good friend's house. Imagine this place in as much detail as you can—the sounds, the smells, and, most of all, what you *feel* like in this place.

4. Now imagine that you'll soon receive a visitor—a warm and kind friend. This is someone who loves you completely and accepts you exactly for who you are. This can be real person like a friend of yours, a beloved grandparent, or a favorite teacher, or it can be a character from a book you've read, a pet, or even a hero or heroine from a comic book or movie.

5. Imagine this being in as much detail as possible, especially how it feels to be in her presence.

6. You have a choice: you can either go out from your safe place to meet your friend or you can invite your friend in. So imagine that you are doing that now.

7. Allow yourself to sit with the person at just the right distance—feeling completely comfortable and safe, completely accepted and loved—remembering that your Compassionate Friend cares deeply about you and just wants you to be happy.

8. Take a moment to enjoy how you feel in the presence of your Compassionate Friend.

9. This being is here with you now and can understand exactly what it's like to be you, exactly where you are in your life right now, and understands precisely what you struggle with.

10. Most of all, this person or being accepts and understands you completely for who you are, perhaps in a way that no one else does.

11. Now this person or being has something important to say to you, something that is just what you *need* to hear right now.

12. See if you can listen closely for the words she wants to share, or, perhaps, you just hear it in your own mind—words that are comforting and supportive.

13. And if no words come, that's okay, too. Just enjoy being in the presence of your Compassionate Friend.

14. And now, maybe you have something to say to this friend. This friend is a very good listener, and completely understands you. Is there anything you'd like to say?

15. Enjoy your friend's good company for a few last moments, and then bid your friend farewell, knowing that you can invite her back whenever you need to.

16. You are now alone in your safe place again. Let yourself savor what just happened, perhaps reflecting on the words you heard.

17. And before this meditation ends, please remember that this Compassionate Friend is a part of *you*. The presence you felt and the words you heard are a deep part of yourself. The comfort and safety that you felt is there within you at all times. Know that you can return to this safe place and to this Compassionate Friend whenever you need to.

18. Bringing your attention back to your breath, gently open your eyes.

Now, write about how you feel now. Again, you can use your own words or any of these: terrible, pretty good, not bad, so-so, feeling great today, happy as can be, depressed, sad, angry, frustrated, terrific, jumping for joy, the worst ever, super, awful, so sad!

_____

On a scale from 1 to 10, with 1 being feeling really awful and 10 feeling really great, what number would you use to rate your mood right now? _____

Is there anything else you want to say about how doing this practice was for you? Remember that your experience is just that—*your* experience, and whatever you feel or think about it is totally okay.

_____

_____

What Teens Say

It was a safe place. I just made up a *cool* place. It had all the things I like in one place. I like mountains and I like log cabins, so I kind of pictured that. It was like a foggy morning and it felt just really safe and calm. So when I just feel kind of stressed, I just go there.

And here's another kind of exercise. This exercise is reflective, and helps you access that Compassionate Friend that we all have inside us, who lives there all the time, but sometimes we don't pay attention to. It also helps us tame the other voice that is sometimes shouting in our ear—the critical, self-judgmental voice. It offers a way of cultivating self-compassion, of being kind to ourselves, so that we can feel good about ourselves no matter what—we don't have to depend on self-esteem and how we perform in life to feel good about ourselves.

try this: Cultivating a Compassionate Voice

If you prefer to listen to rather than read this exercise, please download the audio at http://www.newharbinger.com/39843.

First, consider this:

We all have many parts of ourselves, each with its own voice. This exercise is about acknowledging the critical part of ourselves (which we will call the "Inner Critic") and then finding and acknowledging the compassionate and loving part (which we will call the "Compassionate Voice"). This will help us quiet our critic and cultivate compassion.

Take a moment to think about that inner critical voice (aka the Inner Critic) that you have chattering in your head all the time, saying things like *I did that wrong!* or *I'm an idiot! Why did I do that?* or *I know better than that! Why did I do such a dumb thing?* Do you think that the Inner Critic has a purpose? If so, what would it be? Write your answer here:

_____

_____

Other folks have said that the purpose of the Inner Critic is to motivate them, get them to behave better, avoid further criticism, lower expectations so they don't disappoint themselves, and make others feel better so they like them more. If you think that any of these reasons are true for you, you can write them here. Or maybe your Inner Critic has other purposes. You can write those here, too:

_____

_____

Do you think your Inner Critic might be sticking around to protect you in some way? (Circle one.)

     Yes   No

If so, can you say how the Inner Critic is trying to protect you?

_____

Now think about a behavior that you would like to *change*—something you continue to beat yourself up about. It's important that you don't identify a characteristic you can't change—like having a big nose. (When I was a teen, I thought my nose was really big. I definitely would have chosen to change it if I could.)

Rather, choose a behavior that you think is unhelpful, for example:

"I eat too much."

"I don't exercise enough."

"I procrastinate with my schoolwork."

"I can be irritable and impatient with my parents or siblings."

Now write down the words that you typically say to yourself when you find yourself criticizing this behavior. What does your Inner Critic say, and what tone of voice does it use?

_____

_____

Take a moment to consider how much suffering the Inner Critic has caused you so far. Write about this here:

_____

_____

Try giving yourself compassion for how hard it is to continually hear these harsh words. Write yourself some kind words here, maybe something like "I'm so sorry you have had to hear these words for so long," "You are such a kind, good person, and I know you are trying so hard to do your best—that's what counts," or "You have such a good heart." Remember, like all of us, you are deserving of these kind words.

_____

_____

_____

Now think about this: Do you think your Inner Critic has been trying to *protect* you in some way, even if it's hurting you in the process? Do you think its original intention might have been good?

_____

_____

If so, please write down how your Inner Critic might be trying to keep you safe, protecting you from some perceived risk or danger.

_____

_____

_____

If you can't find any way that the Inner Critic is trying to help you—and sometimes it doesn't—please continue to give yourself compassion for the suffering you've experienced because of all this self-criticism in the past.

You can do that now. Take your time.

You can write more words here. Let it flow—feel free to say lots and lots and lots of kind words to yourself.

_____

_____

_____

If you have identified some way that your Inner Critic might be trying to keep you safe, see if you can acknowledge that effort by your Inner Critic and write down a few words of thanks. Let the Inner Critic know that even though it may not be helping you in the long run, its intention was good, and it was trying its best. You can write those words here:

_____

_____

_____

Now that we've heard from one part of you, the Inner Critic, we're ready to hear from another part of you. This is:

a kinder, gentler voice

a part of you that is wise

a part of you that maybe recognizes that this behavior is causing you some harm

a part of you that also wants you to change, but for very different reasons

Let's see if we can get in touch with this voice.

If it's comfortable for you, put your hands over your heart, feeling the light pressure of your hands and the warmth of them against your heart.

Now think about the issue you are struggling with.

Silently say some words or phrases to yourself. Use words or phrases that you can really relate to, that work for you. Here are some suggestions:

I really care about you, I know how hard it is sometimes being a teen, dealing with all that you're facing right now.

You're experiencing changes in your brain, changes in your body, changes in the way you relate to friends and family, and pressures at school.

It's really, really hard being a teen.

And I don't want you to suffer anymore!

If you have trouble doing this, or trouble thinking of phrases that work for you, think of what a good friend would say to you or what your pet might say to you if your pet could talk. What would your pet say if it knew that you were struggling with something? (Actually, pets often know when we're struggling with something, don't they?) What words would they say?

Maybe something like this:

I care about you and I'm here to protect you and take care of you.

I love you and I want you to be happy.

You're the best—and I'll be here for you always.

Now in the lines below, you can write a letter to yourself, freely and spontaneously, in the voice of your compassionate self, a Compassionate Friend, or a pet.

_____

_____

_____

_____

_____

_____

_____

_____

_____

_____

After you have finished writing your letter, feel free to write about your experience below. Often writing really helps us clarify our thoughts.

_____

_____

_____

_____

_____

What was this exercise like for you? Is there anything that stands out to you about this exercise?

_____

_____

_____

Were you able to hear the Inner Critic? What was that like?

_____

_____

_____

Were you able to hear the Compassionate Voice? What was that like?

_____

_____

_____

If you were able to hear the Compassionate Voice, was it helpful? If so, in what way?

_____

_____

_____

_____

_____

_____

So the Inner Critic, the voice that is judging, critical, and sometimes really harsh, often has had good intentions. It has tried to keep you safe, but maybe it didn't have such great ways to do that, and it came on a little too strong—and, okay, maybe downright nasty. We can acknowledge its good intentions and, at the same time, listen to that other voice that we have inside us—the kind and big-hearted voice that we often use with our friends—and use that voice to be kind to ourselves.

my thoughts

Do you have any thoughts as this chapter comes to a close? What do you think about some of the practices and exercises in this chapter? Did any of these feel particularly "right" to you? You can either write about or draw your reactions here:

conclusion

So what have we learned? Self-compassion is quite different from self-esteem. Self-compassion is a way of relating to ourselves that is there for us always, not just when things are going well. In fact, it's particularly handy to have around when we're feeling bad and we need some kindness and support. And there are lots of ways that we can cultivate our self-compassion. One particularly helpful way is to remember that we have a wise, Compassionate Voice within us always. Through practice, we can learn to listen to that Compassionate Voice more often. And in time, that starts to feel—well, *good*. It's pretty comforting to know that we can actually be a kind, supportive, and loving friend to ourselves!

In the next chapter, we're going to explore what's really important to us—our values—and what happens when we get distracted and don't live according to our values. And, of course, how self-compassion can help. So turn the page!

Part 4

Embracing Who I Am— and Finding Joy!

chapter 10

living by my values

Somewhere deep inside us, we know what's truly important to us. We know what we value. We may not necessarily be living by these values, but deep in our core, we know what we believe in. For this reason, we call these deep-inside values *core values*. And in order to be truly self-compassionate, we need to know what those things are so that we can guide ourselves back when we stray from our core values (which we'll all do from time to time). Often when we're not living by our core values, we tend to be particularly hard on ourselves because we don't like how we're being in the world and acting toward others. At these times, we often don't like ourselves very much, so we really need self-compassion.

For some of us, what we value might be having close family and friends around, for others it might be being out in nature, or following our religious beliefs, or getting a good education. For others, it might be about being a political or humanitarian activist or having our lives filled with art and music. For most of us, it's a number of these things, but in different degrees of importance.

In this chapter, we're going to do an exercise to figure out what our core values are, and then make a promise to ourselves to try to guide ourselves back to those values when we forget. This exercise is called My House/My Self because it uses the image of a house to help us access what it is that is important to us.

try this: My House/My Self

Here are a series of questions to answer about your "house" and an image of a house to help you get in touch with your core values. If you feel like you don't need to answer all the questions to uncover your values, that's okay—they're just to get you thinking. You can write your answers either below or on the house image itself. Make sure you do this exercise when you have at least twenty minutes or so. You want to have time to really reflect on the questions.

Foundation: What are those things that are foundational, or most important, to you?

Window: When you look out the window of your house into the future, what do you see?

Path to Front Door: What are the things that lead you to your house—those things that you believe in?

Garden: What kinds of things would you like to grow and cultivate in your life?

Inside House: Who are those people who have influenced you in your life?

Smoke from Chimney: What parts of yourself would you like to send out into the world?

Fence: What are those things that you want to keep away from your house?

Roof: What keeps you inside? What is limiting you?

Bricks: What holds you together?

If you would like to add something to your house, feel free to do so.

Now that you have an idea of what is most important to you and also the things that sometimes limit you, what do you think are the biggest obstacles to you living by your values?

_____

_____

_____

A lot of teens feel that peer pressure makes it challenging for them to live by their core values. In other words, a teen might want to spend an hour sitting in the woods by herself sketching, but she feels pressure to go to the mall with her friends—it's often hard to say no to friends.

How might peer pressure be preventing you from living by your core values?

_____

_____

_____

_____

Can you be compassionate to yourself right now knowing that it's super hard to say no to friends? What are some kind words that you can say to yourself? Maybe something like *It's hard being a teen, and hard to always know what the right thing is for me. And it's okay if I forget what's really important to me sometimes—I can always come back to living out my values when I remember.*

_____

_____

_____

_____

_____

So rather than criticizing yourself for not living by your core values, you can forgive yourself, reminding yourself that you're human. Then you can start again.

And there are things that we can do to help us remember our core values. One thing is to make a promise to ourselves that when we realize that we've strayed and aren't living by what's really important to us, we'll make an effort to come back to what we value.

The promise that we make is just a guidepost, or a reminder, of where to come back to. We know that inadvertently we'll break that promise because we're human, but then we'll remember—Oh yeah, I have this promise to myself—and we'll return to what we value. That's it!

try this: A Promise to Myself

To make a promise to yourself, first identify a core value that you'd like to keep in your life pretty much every day. Examples of core values might be family, honesty, education, religion, spending time outside, or spending time with friends.

What would that core value be? _____

What is a promise that you can make to live by this core value? For example, if the core value is having art in your life, an example of a promise might be "I promise that I'll take time to sketch every day." Or if a core value is being a good friend, a promise might be "I promise that I will try to really listen fully when my friends are speaking."

What's your promise that you'd like to make to yourself?

_____

_____

_____

One idea is to write this promise down and put it somewhere that you'll see often, like on your mirror, computer, or phone. Another idea is to set the alarm on your phone so that it alerts you at certain times, and every time it does, remind yourself to say your promise to yourself. Or you know all those passwords that you need to use to get into websites? Take a word or two from your promise and use that as your password! That will help remind you!

It doesn't really matter how you do it, the important thing is to remind yourself to come back to what you value, and not to be hard on yourself when you get distracted and forget. So reminding yourself of what your core values are, and refraining from beating yourself up when you stray from your values, will help you to be less self-critical and embrace who you are.

my thoughts

What are ways you can keep your core values up close and center? Do you think this will help you feel like you're being truer to yourself—or be less self-critical? Write or draw your thoughts here:

_____

_____

_____

_____

conclusion

Living by our core values helps us feel more comfortable in our skin, be less self-critical, and simply be who we are. When we forget and stray from what's important to us, we can be kind to ourselves, remembering that we're human, and come back to living by our values. No self-judgment is necessary. This helps us really value and appreciate ourselves.

In the next chapter, you'll read about how you can find things to be grateful for in so many places in your life—and when that happens, you'll not only feel less sad and anxious, but you'll feel—well, appreciative and joyful!

chapter 11

gratitude and self-appreciation

Gratitude—we hear that word a lot, especially around Thanksgiving. And if you're like me, it feels kind of heavy. Like something we should have, but maybe we don't—or, at least, we don't a lot of the time. And then we feel guilty because we know we should be feeling grateful—and well, that feeling just isn't there.

So why don't we feel grateful more?

First of all, know that it's not your fault. Remember negativity bias from chapter 5? From an evolutionary standpoint, we are biologically geared to keep ourselves safe. And in order to do that, we have to keep an eye out for things that could hurt us. So we easily notice things like glances from someone that could be interpreted as hurtful, a comment from a classmate that might be offensive, and even some little mistake you may have made that might hurt you sometime in the future.

We pick up on all the negative stuff so we'll be prepared to protect ourselves. We'll be on guard. But guess what? This often leads to us misinterpreting events as way more negative than they are, and going to great lengths to protect ourselves.

Can you think of a time when you misinterpreted an event as way more negative than it really was? (Circle any that apply.)

Are you kidding? This happens all the time!

Yeah, once.

Are you living in my brain or something? How did you know?!

This never happened to me. Never. And I never tell lies either.

Okay, okay, you made your point. Let's move on!

We are hardwired to keep ourselves "safe," but that doesn't mean we're happy.

So in order to be happy (and grateful), we have to make an extra special effort to look for the good that's out there—for the things that make us smile, the things that bring us joy, even momentarily. And once we start to look—once *you* start to look—you'll see a ton of things like that out there. Prepare to be amazed.

Let's try it. The following activity is a fun way to get you to start noticing all the things in your everyday life that bring you joy.

try this: Phinding Photos with Your Phone (I know, corny on the "phinding," but I couldn't resist!)

This is an activity that you can do in twenty minutes, an hour, or longer. It's super enjoyable, so you're likely to want to take some time to do it. You'll need your cell phone (but if you don't have one, that's okay—you can skip the taking photos part).

1. This is a great activity to do outside, but inside will work, too. As you leisurely stroll along, keep an eye out for things that make you smile. This could be anything. It could be the color of a flower that's in bloom, or your pet curled up asleep on the couch, or an interesting insect that's crawling on the bark of a tree.

2. Spend some time with each thing that you come across, really allowing yourself to enjoy it. Use all your senses—feel the texture of it, maybe notice the smell, and just observe it closely. Remember when you were a kid and spent what seemed like hours watching ants crawl out of an anthill? Kind of like that.

3. Notice how you feel as you pay attention to these things. And take a photo or two of it with your phone.

4. After you've spent some time with one thing, stroll along until you find the next thing that makes you smile. And then do the same with this thing—spend some time enjoying it using your senses, take a couple of photos, and then when you're ready, move on.

5. If you want, when you get back home, download the photos onto your computer so that you have a folder of things that make you smile. Next time you're feeling bad or having a rough day, open the folder and voilà! A whole array of things that bring you joy!

Write or draw how you feel now after doing this activity. Do you feel differently than you did before you began?

_____

_____

_____

How might doing this activity on a regular basis help you feel better in general? (Check any that apply.)

☐ It would help me get outside my self-critical head.

☐ It would help remind me what really makes me happy.

☐ It would help me get my phone back from my parents when they take it away from me.

☐ Strolling sounds like it's for old people.

☐ Strolling sounds like it's for toddlers—like you should be in a stroller.

☐ It wouldn't improve things much—I'm already aware of things that make me happy!

☐ This activity made me love the world and everything in it.

☐ Anything else? _____

As you've probably figured out, this activity helps you feel happier in general by correcting your "negativity bias." It turns your focus from yourself and your struggles to the things that bring you joy.

And the super-great realization here is that these things are available to us in every moment, but we usually don't see them because we're so busy romping around in our self-critical head.

A variation of this activity is to keep your phone with you during the day (I know, you do this already!), and as you're going through your day, notice things that make you smile. It might be an event—hanging out with friends, your team winning a game, or a favorite band that you saw in concert—or some brief moment—a goofy expression on your little sister's face or a friend doing something outrageous on a skateboard. Take photos, and keep them on your phone or in a folder on your computer. Don't forget to visit them often!

What Teens Say

I've been appreciating nature a lot more—the sounds and the way it feels and how beautiful it is and how nature's like the one thing we can't control.

I don't know—it's made me realize how big and beautiful this world is, and no matter how bad your life gets, there's still like a beautiful world out there.

Now that you've noticed all these things around you that bring you joy, it's just a half step away to feeling grateful. In fact, I bet you're finding that you already feel grateful!

The next exercise is an eye-opener in seeing how feeling grateful leads to feeling better overall.

Notice how you're feeling right now, and mark it on this line with an X. (Actually, why do we always mark with an X? Mark it with whatever mark you'd like!)

|_____|

In the dumps.   Joyful!

Now we're going to do another brief exercise.

try this: What Am I Grateful For?

What you'll need: the stopwatch on your phone or some kind of timer.

For one minute, you'll write down all the things that you're grateful for. Some things might be obvious, like friends and family and a house to live in, but don't forget the little things—like your favorite pen, the smell of bacon cooking, or gourmet multiflavored jelly beans.

During the minute, just keep writing! If you get stuck and can't think of something, just write the same thing over and over until something new pops into your head.

Set the timer for one minute—and start writing!

After a minute, rate how you feel—with whatever mark you'd like.

|_____|

In the dumps.   Joyful!

Feel any better?

Usually when we take time to notice those things in our lives that we're grateful for, we feel better. In fact, research has shown that those who make a point to notice the positive things in their lives generally feel better than those who don't.

You can make a habit of doing this—noticing what you're grateful for. You can do this by keeping a daily gratitude journal, either in a regular paper journal or on a gratitude app—there are a bunch that you can download and use from your phone. Try it—it's actually kind of fun!

What Teens Say

I realize now that there are other things besides stress, and you know, you can think about other things going on, like abstract ideas and things.

I want to do better in the world if that makes sense. I'm less focused on all the things that I'm doing wrong and more focused on what I **could** be doing better.

So making a point to notice the things that we are genuinely grateful for allows us to focus less on the negative—like the kinds of things we beat ourselves up about—and focus our energy on all the other things that are out there in the world. As our teens above said, they could focus on abstract ideas and doing better in the world.

In addition to making a point to appreciate all the good things in our lives, we can also make a point to notice the good things in *ourselves*.

What? Appreciate ourselves? Are you kidding me? Now you're going too far . . .

No, I'm not, really.

I know that appreciating ourselves usually seems like a foreign concept to us, especially when we're teens. But my guess is that, by this point in the book, you've realized that we're not kidding—that it is genuinely, honestly, absolutely possible to really appreciate yourself, to really value who you are, *especially* when you mess up and things aren't going so well.

It just takes practice.

And practice means reminding ourselves in those moments that we're being self-critical and that we don't have to be. That there's another way. That by being self-compassionate, we can learn to really appreciate ourselves.

We tend to remember so clearly the things that we *don't* like about ourselves, but not so much the things that we *do* like about ourselves. Remember negativity bias? It's here, too.

So here's a self-appreciation practice that you can do at any time. It will help you remember those qualities that you have that you genuinely like about yourself. Good-bye, self-criticism—hello, self-appreciation!

try this: Appreciating Me!

1. Get comfortable, close your eyes, and notice any sensations you might be feeling in your body. Take a few minutes to do this.

2. Now take a moment to think about one or two things that you genuinely appreciate about yourself, that you really, deep down like about yourself. No one is listening to your thoughts, so you can be completely honest with yourself. And it doesn't have to be a big thing. It can be something little. It all counts, big and little. Take all the time you need to do this. No rush.

3. Sometimes it's easier to appreciate ourselves when we remember that much of what is good in us and what we enjoy about ourselves comes from what others have given us in so many ways throughout our lives.

4. Can you think of any people who helped you develop your good qualities? Maybe friends, parents, or teachers? Maybe even authors of books who had a positive impact on you? As each one comes to mind, you can send them some gratitude or appreciation.

5. When we honor ourselves, we honor those who have helped and supported us all through our lives.

6. Let yourself savor, just for this moment, feeling really good about yourself. Dwell in it. Enjoy it. Let it soak in. Remember, nobody is watching.

7. And keep in mind that you don't have to be the best or to be perfect in order to appreciate something about yourself. It can be just one thing. And it doesn't have to be a huge thing.

8. When you're ready, you can open your eyes.

Notice how you feel right now, and see if you can find any gratitude or appreciation words in the puzzle below. Hint: There are eleven words or phrases. See if you can find them all! (You can find the answers online at http://www.newharbinger.com /39843.)

O	T	R	H	J	X	J	N	L	S	G	I	L	V	M
P	P	E	M	A	F	G	E	H	Q	Z	I	M	D	Y
T	C	V	C	S	P	D	R	I	L	K	A	X	E	Y
O	B	E	Z	I	P	P	L	A	E	I	Z	K	O	S
G	F	N	Y	T	F	K	Y	A	Q	P	I	J	J	U
N	S	A	V	N	K	I	N	T	E	A	H	S	R	P
I	Q	H	K	N	Q	E	R	M	O	T	Y	I	S	E
Z	O	T	B	K	W	E	O	R	I	B	E	F	G	R
A	Y	R	F	M	N	S	P	W	E	R	E	C	X	G
M	N	E	E	W	E	Y	D	K	D	T	A	M	S	R
A	J	T	I	W	G	E	G	O	O	D	T	G	E	E
X	Z	T	A	J	L	D	A	B	O	S	T	O	N	A
K	X	E	Y	L	V	F	D	N	Q	O	B	E	X	T

Although many of us have been taught that it's wrong or self-serving to think about our good qualities, when we take time to value ourselves, we often feel more comfortable and open to valuing others. So there's nothing wrong with a little self-appreciation!

What Teens Say

I guess I don't worry now so much about others liking me—because I like me!

my thoughts

Write any thoughts you have about this chapter. Did anything surprise you? Can you foresee making it a habit to be grateful? Remember, you can also draw instead of writing here:

conclusion of conclusions

Although we know deep inside that we are grateful for many things in our lives, we often don't take the time to notice what we're grateful for. When we do, we often experience a new level of contentment and ease, and that nagging self-critical voice is silent for a few moments. Taking time to really appreciate the qualities that we like about ourselves allows us to be okay with just who we are, acknowledging that we're not perfect and knowing that "not-perfect" is perfectly okay. Perfectly imperfect beings, we can embrace who we are, exactly as we are. Just for this one moment. And the next...and the next...

acknowledgments

This book would not be possible without the help of many. Most importantly, I'd like to acknowledge Lorraine Hobbs, my colleague, collaborator, and cherished friend. After much joy and some labor pains, we gave birth to *Making Friends with Yourself,* which established the foundation for this book. As *Making Friends with Yourself* served as the foundation, Kristin Neff and Chris Germer provided the bedrock of this book through their development of Mindful Self-Compassion, the adult course from which *Making Friends with Yourself* is adapted. Many of the activities, meditations, and exercises in this book have been modified from the program they created. They are the true "self-compassion" pioneers, and my mentors. I have enormous gratitude for their vision, wisdom, and fortitude to create a program that has benefited so many people, and continues to spread internationally. Steve Hickman has supported me throughout my Mindful Self-Compassion teacher training, and was always encouraging in the creation of the teen program. Michelle Becker has been a mentor, friend, and inspiration in truly living compassion. These individuals have provided a web of support and guidance for me throughout my journey learning about, living, and teaching self-compassion.

I am also tremendously thankful to Kate Murphy, who created some of the artwork in this book, my illustrator Zanne DeJanvier, and my editors at New Harbinger, Tesilya Hanauer and Vicraj Gill. Their support provided me with the gentle guidance that I needed as I traversed the writing of this book. Susan Girdler, my research mentor and friend, provided me with a quiet office in which I wrote most of this book, and the researchers in her lab provided a positive and uplifting environment that made writing this book even more enjoyable. I'd also like to acknowledge Trish Broderick, who took me under her wing as a graduate student in search of a mindfulness mentor and who has offered wisdom and guidance since then. Mary Jane Moran and Priscilla Blanton, my graduate school professors and friends, believed in me throughout my graduate school career, even when mindfulness and self-compassion were virtually unknown in east Tennessee, and have continued to encourage and support my path since then.

My parents, David and Doris Bluth, have given me so much support throughout my life, and so often when I needed it most. And, of course, none of this would be possible without the constant support of Dale, my *b'shareta,* who is my emotional home and safe haven, and who has taught me what it truly means to feel protected and safe.

In memoriam, I must recognize my maternal grandfather, Morris Finkelstein, who always seems to show up as my Compassionate Friend. Papa Morris is my internalized voice of wisdom and compassion. I was lucky to share my childhood with him, and so grateful that he is here with me throughout my adulthood as well. May all youth feel as loved, safe, and accepted as I did in the presence of Papa Morris.

Karen Bluth, PhD, earned her doctoral degree in child and family studies at the University of Tennessee. She is currently research faculty in the Program on Integrative Medicine in the Department of Physical Medicine and Rehabilitation at the University of North Carolina School of Medicine. Her work focuses on the roles that mindfulness and self-compassion play in promoting well-being in teens. Bluth was awarded a Francisco J. Varela research award from the Mind and Life Institute in 2012, which allowed her to explore the effects of a mindfulness intervention on adolescents' well-being through examining stress biomarkers. In spring 2015, she received internal University of North Carolina funding to explore relationships among mindfulness, self-compassion, and emotional well-being in teens in grades 7–12. With current NIH funding, she is part of a research team at the University of North Carolina that is studying the teen adaptation of Kristin Neff and Christopher Germer's *Mindful Self-Compassion* program.

In addition to her research, Bluth regularly teaches mindfulness and mindful self-compassion courses to both adults and teens in the Chapel Hill, NC, area and regularly gives talks and leads workshops at schools and universities. In collaboration with Lorraine Hobbs, Bluth has adapted Kristin Neff and Christopher Germer's *Mindful Self-Compassion* program for an adolescent population. A former educator with eighteen years classroom experience, Bluth is currently associate editor of the academic journal *Mindfulness*.

Foreword writer **Kristin Neff, PhD**, is currently associate professor of educational psychology at the University of Texas at Austin. She is a pioneer in the field of self-compassion research, conducting the first empirical studies on self-compassion over a decade ago. In addition to writing numerous academic articles and book chapters on the topic, she is author of the book *Self-Compassion*, released by William Morrow. In conjunction with her colleague Christopher Germer, she developed an empirically supported eight-week training program called *Mindful Self-Compassion*, and offers workshops on self-compassion worldwide. Neff is also featured in the best-selling book and award-winning documentary *The Horse Boy*, which chronicles her family's journey to Mongolia, where they trekked on horseback to find healing for her autistic son.

More ⏱ Instant Help Books for Teens

An Imprint of New Harbinger Publications

STUFF THAT SUCKS

A Teen's Guide to Accepting What
You Can't Change & Committing
to What You Can

ISBN: 978-1626258655 / US $12.95

**MINDFULNESS FOR TEEN
DEPRESSION**

A Workbook for Improving
Your Mood

ISBN: 978-1626253827 / US $16.95

**THE BODY IMAGE
WORKBOOK FOR TEENS**

Activities to Help Girls Develop
a Healthy Body Image in an
Image-Obsessed World

ISBN: 978-1626250185 / US $16.95

**THE PERFECTIONISM
WORKBOOK FOR TEENS**

Activities to Help You Reduce
Anxiety & Get Things Done

ISBN: 978-1626254541 / US $16.95

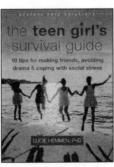

**THE TEEN GIRL'S
SURVIVAL GUIDE**

Ten Tips for Making Friends,
Avoiding Drama & Coping
with Social Stress

ISBN: 978-1626253063 / US $16.95

**THINK CONFIDENT,
BE CONFIDENT FOR TEENS**

A Cognitive Therapy Guide to
Overcoming Self-Doubt & Creating
Unshakable Self-Esteem

ISBN: 978-1608821136 / US $16.95

Register your **new harbinger** titles for additional benefits!

When you register your **new harbinger** title—purchased in any format, from any source—you get access to benefits like the following:

- Downloadable accessories like printable worksheets and extra content

- Instructional videos and audio files

- Information about updates, corrections, and new editions

Not every title has accessories, but we're adding new material all the time.

Access free accessories in 3 easy steps:

1. Sign in at NewHarbinger.com (or **register** to create an account).

2. Click on **register a book**. Search for your title and click the **register** button when it appears.

3. Click on the **book cover or title** to go to its details page. Click on **accessories** to view and access files.

That's all there is to it!

If you need help, visit:

NewHarbinger.com/accessories

new harbinger
CELEBRATING
40 YEARS